MEDIA COVERAGE OF TERRORISM
AND
METHODS OF DIFFUSION

MEDIA COVERAGE OF TERRORISM
AND
METHODS OF DIFFUSION

Rahul Chakravarti

MD PUBLICATIONS PVT LTD
NEW DELHI
www.mdppl.com

Published by :

MD Publications Pvt Ltd
"MD House", 11, Darya Ganj,
New Delhi - 110 002
Phone :   +91-11-41563325 (Customer Service)
          +91-11-41562846 (Editorial)
E-mail (orders): order@mdppl.com
Website: www.mdppl.com

ISBN: 978-81-7533-182-2

Published and Printed by Mr. Pranav Gupta on behalf of **MD Publications Pvt Ltd** at Times Press, New Delhi.

PREFACE

The terrorist situations represent a very rare circumstance in which straightforward reporting about goals, aspirations, and demands of terrorists may be socially irresponsible. Although it concedes that international news reporting about terrorist incidents often reflects self-restraint, but during past 40 years there have been many instances of media coverage of terrorist events as problematic and irresponsible, evoking public criticism and antagonising authorities. In the countries like Canada, Germany, Israel, U.S., and the U.K., it is urged that there is the need to develop a set of guidelines for responsible media coverage of terrorism.

Terrorists, governments, and the media see the function, roles and responsibilities of the media, when covering terrorist events, from differing and often opposing perspectives. Such perceptions drive respective behaviors during terrorist incidents—often resulting in tactical and strategic gains, or losses, to the terrorist operation and the overall terrorist cause. The challenge to the governmental and press community is to understand the dynamics of terrorist enterprise and to develop policy options to serve government, media and societal interests.

Terrorists must have publicity in some form if they are to gain attention, inspire fear and respect, and secure favorable understanding of their cause, if not their act. Governments need public understanding, cooperation, restraint, and loyalty in efforts to limit terrorist harm to society and in efforts to punish or apprehend those responsible for terrorist acts. Journalists and the media in general pursue the freedom to cover events and issues without restraint, especially governmental restraint.

The book is written in such impressive manner which questions whether national and international security is more important than freedom of information. It explores the role of television news and some issues regarding how concessions to terrorist demands result from press publicity also are left unresolved and recommended for serious collections in journalism ethics and freedom of information.

I am very grateful to Mr. Pranav Gupta, Director of MD Publications Pvt Ltd, New Delhi, for his cooperation and support for this book.

Rahul Chakravarti

CONTENTS

1

Introduction

"Terrorism" is a term that cannot be given a stable definition. Or rather, it can, but to do so forestalls any attempt to examine the major feature of its relation to television in the contemporary world. As the central public arena for organising ways of picturing and talking about social and political life, TV plays a pivotal role in the contest between competing definitions, accounts and explanations of terrorism. Politicians frequently try to limit the terms of this competition by asserting the primacy of their preferred versions. Jeanne Kirkpatrick, former U.S. representative to the United Nations, for example, had no difficulty recognising "terrorism" when she saw it, arguing that "what the terrorist does is kill, maim, kidnap, torture. His victims may be schoolchildren.... industrialists returning home from work, political leaders or diplomats". Television journalists, in contrast, prefer to work with less elastic definitions. The BBC's News Guide for example, advises reporters that "the best general rule" is to use the term "terrorist" when civilians are attacked and "guerrillas" when the targets are members of the official security forces.

Which term is used in any particular context is inextricably tied to judgements about the legitimacy of the action in question and of the political system against which it is directed. Terms like "guerrilla", "partisan" or "freedom fighter" carry positive connotations of a justified struggle against an occupying power or an oppressive state; to label an action as "terrorist" is to consign it to illegitimacy. For most of the television age, from the end of World War II to the collapse of the Soviet Union, the deployment of positive and negative political labels was an integral part of Cold War politics and its dualistic view of the world. "Terrorism" was used extensively to characterise enemies of the United States and its allies, as in President Reagan's assertion in 1985, that Libya, Cuba, Nicaragua and North Korea constituted a "confederation of terrorist states" intent on undermining American attempts "to bring stable and democratic government" to the developing world.

Conversely, "friendly" states, like Argentina, could wage a full scale internal war against "terrorism", using a definition elastic enough to embrace almost anyone who criticised the regime or held unacceptable opinions, and attract comparatively little censure despite the fact that this wholesale use of state terror killed and maimed many more civilians than the more publicized incidents of "retail" terror — assassinations, kidnappings and bombings. The relations between internal terrorism and the state raise particularly difficult questions for liberal democracies. By undermining the state's claim to a legitimate monopoly of force within its borders, acts of "retail" terror pose a clear threat to internal security. And, in the case of sub national and separatist movements which refuse to recognise the integrity of those borders, they directly challenge its political legitimacy.

Faced with these challenges, liberal democracies have two choices. Either they can abide by their own declared principles, permit open political debate on the underlying causes and claims of terrorist movements, uphold the rule of law, and respond to insurgent violence through the procedures of due process. Or they can curtail public debate and civil liberties in the name of effective security. The British state's response to the conflict in Northern Ireland,

and to British television's attempts to cover it, illustrate this tension particularly well. Television journalism in Britain has faced a particular problem in reporting "the Irish Question" since the Republican movement has adopted a dual strategy using both the ballot box and the bullet, pursuing its claim for the ultimate reunification of Ireland electorally, through the legal political party, Sinn Fein, and militarily, through the campaign waged by the illegal Irish Republican Army. Added to which, the British state's response has been ambiguous.

Ostensibly, as Prime Minister Thatcher argued in 1990, although "they are at war with us" "we can only fight them with the civil law." Then Home Secretary, Douglas Hurd, admitted in 1989 that, in his view "with the Provisional IRA...it is nothing to do with a political cause any more. They are professional killers....No political solution will cope with that. They just have to be extirpated". Television journalists' attempts to explore these contradictions produced two of the bitterest peacetime confrontations between British broadcasters and the British state. Soon after British troops were first sent to Northern Ireland in the early 1970s, there were suspicions that the due process of arrest and trial was being breached by a covert but officially sanctioned shoot-to-kill campaign against suspected members of Republican paramilitary groups.

In 1988, three members of an IRA active service unit were shot dead by members of an elite British counter terrorist unit in Gibraltar. Contrary to the initial official statements, they were later found to be unarmed and not in the process of planting a car bomb as first claimed. One of the leading commercial television companies, Thames Television, produced a documentary entitled *Death on the Rock*, raising questions about the incident. It was greeted with a barrage of hostile criticism from leading Conservative politicians, including Prime Minister Thatcher. The tone of official condemnation was perfectly caught in an editorial headline in the country's best-selling daily paper *The Sun* claiming that the programme was "just IRA propoganda."

The representation of the Provisional IRA was at the heart of the second major conflict, over a BBC documentary entitled *At the Edge of the Union*. This featured an extended profile of Martin McGuiness of Sinn Fein, widely thought to also be a leading IRA executive responsible for planning bombings. The programme gave him space to explain his views and showed him in his local community and at home with his family. The then Home Secretary, Leon Brittan, (who had not seen the film) wrote to the Chairman of the BBC's Board of Governors urging them not to show it, arguing that "Even if [it] and any surrounding material were, as a whole, to present terrorist organisations in a wholly unfavourable light, I would still ask you not to permit it to be broadcast".

The Governors convened an emergency meeting and decided to cancel the scheduled screening. This very public vote of no confidence in the judgement of the corporation's senior editors and managers was unprecedented and was met with an equally unprecedented response from BBC journalists. They staged a one-day strike protesting against government interference with the Corporation's independence. In his letter, Brittan had claimed that it was "damaging to security and therefore to the public interest to provide a boost to the morale of the terrorists and their apologists in this way". Refusing this conflation of "security" with the "public interest" is at the heart of television journalism's struggle to provide an adequate information base for a mature democracy.

As the BBC's Assistant Director General put it in 1988, "It is necessary for the maintenance of democracy that unpopular, even dangerous, views are heard and thoroughly understood. The argument about the 'national interest' demanding censorship of such voices is glib and intrinsically dangerous. Who determines the 'national interest?' How far does the 'national interest' extend?" His argument was soundly rejected by the government. In the autumn of 1988, they instructed broadcasters not to transmit direct speech from members of eleven Irish organisations, including Sinn Fein. This ban has since been lifted, but its imposition illustrates the permanent potential for conflict between official conceptions of security and the national interest and broadcasters' desire to provide full information, rational debate and relevant contextualisation on areas of political controversy and dispute.

As the BBC's former director general, Ian Trethowan, pointed out, the basic dilemma posed by television's treatment of terrorism is absolutely "central to the ordering of a civilised society: how to avoid encouraging terrorism and violence while keeping a free and democratic people properly informed." Television's ability to strike this balance is not just a question for news, current affairs and documentary production however. The images and accounts of terrorism offered by television fiction and entertainment are also important in orchestrating the continual contest between the discourse of government and the state, the discourses of legitimated opposition groups, and the discourses of insurgent movements. This struggle is not simply for visibility—to be seen and heard. It is also for credibility—to have one's views discussed seriously and one's case examined with care. The communicative weapons in this battle are unevenly distributed however.

As the saturation coverage that the U.S. news media gave to the Shi'ite hijacking of a TWA passenger jet at Beirut in 1985 demonstrated very clearly, spectacular acts of retail terror can command a high degree of visibility. But the power to contextualise and to grant or with old legitimacy lies with the array of official spokespeople who comment on the event and help construct its public meaning. As the American political scientist, David Paletz, has noted, because television news "generally ignores the motivations, objectives and long-term goals of violent organisations" it effectively prevents "their causes from gaining legitimacy with the public". This has led some commentators to speculate that exclusion from the general process of meaning making is likely to generate ever more spectacular acts designed to capitalise on the access provided by the highly visible propaganda of the deed.

Bernard Lewis, one of America's leading experts on the Arab world noted in his comments on the hijacking of the TWA airliner, that those who plotted the incident "knew that they could count on the American press and television to provide them with unlimited publicity and perhaps even some form of advocacy," but because the coverage ignored the political roots of the action in the complex power struggles within Shi'ite Islam, it did little to explain its causes or to foster informed debate on appropriate responses. As the television critic of the *Financial Times* of London, put it; "There is a criticism to be made of the coverage of these events, but it is not that television aided and abetted terrorists. On the contrary, it is that television failed to convey, or even to consider, the reasons for what President Reagan called 'ugly , vicious, evil terrorism.'"

News is a relatively closed form of television programming. It privileges the views of spokespeople for governments and state agencies and generally organises stories to converge around officially sanctioned resolutions. Other programme forms, documentaries for example, are potentially at least, more open. They may allow a broader spectrum of perspectives into play, including those that voice alternative or oppositional viewpoints, they may stage debates and pose awkward questions rather than offering familiar answers. Television in a democratic society requires the greatest possible diversity of open programme forms if it is to address the issues raised by terrorism in the complexity they merit. Whether the emerging forces of technological change, in production and reception, channel proliferation, increased competition for audiences and transnational distribution, will advance or block this ideal is a question well worth examining.

Terrorist Profiling

The isolation of attributes or traits shared by terrorists is a formidable task because there are probably as many variations among terrorists as there may be similarities. Efforts by scholars to create a profile of a "typical" terrorist have had mixed success, if any, and the assumption that there is such a profile has not been proven. Post (1985:103) note that "behavioral scientists attempting to understand the psychology of individuals drawn to this violent political behavior have not succeeded in identifying a unique "terrorist mindset." People who have joined terrorist groups have come from a wide range of cultures, nationalities, and ideological causes, all strata of society, and diverse professions.

Their personalities and characteristics are as diverse as those of people in the general population. There seems to be general agreement among psychologists that there is no particular psychological attribute that can be used to describe the terrorist or any "personality" that is distinctive of terrorists. Some terrorism experts are skeptical about terrorist profiling. For example, Laqueur (1997:129) holds that the search for a "terrorist personality" is a fruitless one. Paul Wilkinson (1997:193) maintains that "We already know enough about terrorist behavior to discount the crude hypothesis of a 'terrorist personality' or 'phenotype.'

The U.S. Secret Service once watched for people who fit the popular profile of dangerousness—the lunatic, the loner, the threatener, the hater. That profile, however, was shattered by the assassins themselves. In interviews with assassins in prisons, hospitals, and homes, the Secret Service learned an important lesson—to discard stereotypes. Killers are not necessarily mentally ill, socially isolated, or even male. Now the Secret Service looks for patterns of motive and behavior in potential presidential assassins.

The same research methodology applies to potential terrorists. Assassins, like terrorists in general, use common techniques. For example, the terrorist would not necessarily threaten to assassinate a politician in advance, for to do so would make it more difficult to carry out the deed. In its detailed study of 83 people who tried to kill a public official or a celebrity in the United States in the past 50 years, the Secret Service found that not one assassin had made a threat. Imprisoned assassins told the Secret Service that a threat would keep them from succeeding, so why would they threaten? This was the second important lesson learned from the study.

The diversity of terrorist groups, each with members of widely divergent national and socio-cultural backgrounds, contexts, and goals, underscores the hazards of making generalizations and developing a profile of members of individual groups or of terrorists in general. Post cautions that efforts to provide an overall "terrorist profile" are misleading: "There are nearly as many variants of personality who become involved in terrorist pursuits as there are variants of personality."

Many theories are based on the assumption that the terrorist has an "abnormal" personality with clearly identifiable character traits that can be explained adequately with insights from psychology and psychiatry. Based on his work with various West German terrorists, one German psychologist, L. Sullwold (1981), divided terrorist leaders into two broad classes of personality traits: the extrovert and the hostile neurotic, or one having the syndrome of neurotic hostility. Extroverts are unstable, uninhibited, inconsiderate, self-interested, and unemotional—thrill seekers with little regard for the consequences of their actions. Hostile neurotics share many features of the paranoid personality they are intolerant of criticism, suspicious, aggressive, and defensive, as well as extremely sensitive to external hostility.

Sullwold also distinguishes between leaders and followers, in that leaders are more likely to be people who combine a lack of scruples with extreme self-assurance; they often lead by frightening or pressuring their followers. Some researchers have created psychological profiles of terrorists by using data provided by former terrorists who became informants, changed their political allegiance, or were captured. Franco Ferracuti conducted one such study of the Red Brigade terrorists in Italy. He analyzed the career and personalities of arrested terrorists by collecting information on demographic variables and by applying psychological tests to construct a typology of terrorists.

Like Post, Ferracuti also found, for the most part, the absence of psychopathology (see Glossary), and he observed similar personality characteristics, that is, a basic division between extroverts and hostile neurotics. By reading and studying terrorist literature, such as group communiqués, news media interviews, and memoirs of former members, it would also be possible to ascertain certain vulnerabilities within the group by pinpointing its sensitivities, internal disagreements, and moral weaknesses. This kind of information would assist in developing a psychological profile of the group.

Post points out that the social dynamics of the "anarchic-ideologues," such as the RAF, differ strikingly from the "nationalist-separatists," such as ETA or the Armenian Secret Army for the Liberation of Armenia (ASALA). From studies of terrorists, Post (1990) has observed indications that terrorists, such as those of the ETA, who pursue a conservative goal, such as freedom for the Basque people, have been reared in more traditional, intact, conservative families, whereas anarchistic and left-wing terrorists (such as members of the Meinhof Gang/RAF) come from less conventional, non intact families.

In developing this dichotomy between separatists and anarchists, Post draws on Robert Clark's studies of the social backgrounds of the separatist terrorists of the ETA. Clark also found that ETA terrorists are not alienated and psychologically distressed. Rather, they are psychologically healthy people who are strongly supported by their families and ethnic community. Post bases his observations of anarchists on a broad-cased investigation of the social background and psychology of 250 terrorists (227 left-wing and 23 right-wing) conducted by a consortium of West German social scientists under the sponsorship of the Ministry of Interior and published in four volumes in 1981-84.

According to these West German analyses of RAF and June Second Movement terrorists, some 25 percent of the leftist terrorists had lost one or both parents by the age of fourteen and 79 percent reported severe conflict with other people, especially with parents (33 percent). The German authors conclude in general that the 250 terrorist lives demonstrated a pattern of failure both educationally and vocationally. Post concludes that "nationalist-separatist" terrorists such as the ETA are loyal to parents who are disloyal to their regime, whereas "anarchic-ideologues" are disloyal to their parents' generation, which is identified with the establishment.

Profiles of terrorists have included a profile constructed by Charles A. Russell and Bowman H. Miller (1977), which has been widely mentioned in terrorism-related studies, despite its limitations, and another study that involved systematically analyzing biographical and social data on about 250 German terrorists, both left-wing and right-right. Russell and Bowman attempt to draw a sociological portrait or profile of the modern urban terrorist based on a compilation and analysis of more than 350 individual terrorist cadres and leaders from Argentinian, Brazilian, German, Iranian, Irish, Italian, Japanese, Palestinian, Spanish, Turkish, and Uruguayan terrorist groups active during the 1966-76 period, the first decade of the contemporary terrorist era. Russell and Bowman (1977:31) conclude:

In summation, one can draw a general composite picture into which fit the great majority of those terrorists from the eighteen urban guerrilla groups examined here. To this point, they have been largely single males aged 22 to 24...who have some university education, if not a college degree. The female terrorists, except for the West German groups and an occasional leading figure in the JRA and PFLP, are preoccupied with support rather than operational roles....Whether having turned to terrorism as a university student or only later, most were provided an anarchist or Marxist world view, as well as recruited into terrorist operations while in the university.

Russell and Miller's profile tends to substantiate some widely reported sociological characteristics of terrorists in the 1970s, such as the youth of most terrorists. Of particular interest is their finding that urban terrorists have largely urban origins and that many terrorist cadres have predominantly middle-class or even upper-class backgrounds and are well educated, with many having university degrees. However, like most such profiles that are based largely on secondary sources, such as newspaper articles and academic studies, the Russell and Miller profile cannot be regarded as definitive.

Furthermore, their methodological approach lacks validity. It is fallacious to assume that one can compare characteristics of members of numerous terrorist groups in various regions of the world and then make generalizations about these traits. For example, the authors' conclusion that terrorists are largely single young males from urban, middle-class or upper-middle-class backgrounds with some university education would not accurately describe many members of terrorist groups operating in the 1990s.

The rank and file of Latin American groups such as the FARC and Shining Path, Middle Eastern groups such as the Armed Islamic Group (Group Islamique Armé—GIA), Hamas, and Hizballah, Asian groups such as the LTTE, and Irish groups such as the IRA are poorly educated. Although the Russell and Miller profile is dated, it can still be used as a basic guide for making some generalizations about typical personal attributes of terrorists, in combination with other information. Edgar O'Ballance (1979) suggests the following essential characteristics of the "successful" terrorist: dedication, including absolute obedience to the leader of the movement; personal bravery; a lack of feelings of pity or remorse even though victims are likely to include innocent men, women, and children; a fairly high standard of intelligence, for a terrorist must collect and analyze information, devise and implement complex plans, and evade police and security forces; a fairly high degree of sophistication, in order to be able to blend into the first-class section on airliners, stay at first-class hotels, and mix inconspicuously with the international executive set; and be a reasonably good educational background and possession of a fair share of general knowledge (a university degree is almost mandatory), including being able to speak English as well as one other major language.

Increasingly, terrorist groups are recruiting members who possess a high degree of intellectualism and idealism, are highly educated, and are well trained in a legitimate profession. However, this may not necessarily be the case with the younger, lower ranks of large guerrilla/terrorist organizations in less-developed countries, such as the FARC, the PKK, the LTTE, and Arab groups, as well as with some of the leaders of these groups.

Russell and Miller found that the average age of an active terrorist member (as opposed to a leader) was between 22 and 25, except for Palestinian, German, and Japanese terrorists, who were between 20 and 25 years old. Another source explains that the first generation of RAF terrorists went underground at approximately 22 to 23 years of age, and that the average age shifted to 28 to 30 years for second-generation terrorists (June Second Movement). In summarizing the literature about international terrorists in the 1980s, Taylor (1988) characterizes their demography as being in their early twenties and unmarried, but he notes that there is considerable variability from group to group.

Age trends for members of many terrorist groups were dropping in the 1980s, with various groups, such as the LTTE, having many members in the 16 to 17 year-old age level and even members who were preteens. Laqueur notes that Arab and Iranian groups tend to use boys aged 14 to 15 for dangerous missions, in part because they are less likely to question instructions and in part because they are less likely to attract attention. In many countries wracked by ethnic, political, or religious violence in the developing world, such as Algeria, Colombia, and Sri Lanka, new members of terrorist organizations are recruited at younger and younger ages.

Adolescents and preteens in these countries are often receptive to terrorist recruitment because they have witnessed killings first-hand and thus see violence as the only way to deal with grievances and problems. In general, terrorist leaders tend to be much older. Brazil's Carlos Marighella, considered to be the leading theoretician of urban terrorism, was 58 at the time of his violent death on November 6, 1969. Mario Santucho, leader of Argentina's People's Revolutionary Army (ERP), was 40 at the time of his violent death in July 1976. Raúl Sendic, leader of the Uruguayan Tupamaros, was 42 when his group began operating in the late 1960s. Renato Curcio, leader of the Italian Red Brigades, was 35 at the time of his arrest in early 1976. Leaders of the Baader-Meinhof Gang were in their 30s or 40s. Palestinian terrorist leaders are often in their 40s or 50s.

Terrorists in general have more than average education, and very few Western terrorists are uneducated or illiterate. Russell and Miller found that about two-thirds of terrorist group members had some form of university training. The occupations of terrorist recruits have varied widely, and there does not appear to be any occupation in particular that produces terrorists, other than the ranks of the unemployed and students. Between 50 and 70 percent of the younger members of Latin American urban terrorist groups were students. The Free University of Berlin was a particularly fertile recruiting ground for Germany's June Second Movement and Baader-Meinhof Gang.

Highly educated recruits were normally given leadership positions, whether at the cell level or national level. The occupations of terrorist leaders have likewise varied. Older members and leaders frequently were professionals such as doctors, bankers, lawyers, engineers, journalists, university professors, and mid-level government executives. Marighella was a politician and former congressman. The PFLP's George Habash was a medical doctor. The PLO's Yasir Arafat was a graduate engineer. Mario Santucho was an economist. Raúl Sendic and the Baader-Meinhof's Horst Mahler were lawyers. Urika Meinhof was a journalist. The RAF and Red Brigades were composed almost exclusively of disenchanted intellectuals.

It may be somewhat misleading to regard terrorists in general as former professionals. Many terrorists who have been able to remain anonymous probably continue to practice their legitimate professions and moonlight as terrorists only when they receive instructions to carry out a mission. This may be more true about separatist organizations, such as the ETA and IRA, whose members are integrated into their communities, than about members of anarchist groups, such as the former Baader-Meinhof Gang, who are more likely to be on wanted posters, on the run, and too stressed to be able to function in a normal day-time job. In response to police infiltration, the ETA, for example, instituted a system of "sleeping commandos."

These passive ETA members, both men and women, lead seemingly normal lives, with regular jobs, but after work they are trained for specific ETA missions. Usually unaware of each others' real identities, they receive coded instructions from an anonymous source. After carrying out their assigned actions, they resume their normal lives. Whereas terrorism for anarchistic groups such as the RAF and Red Brigades was a full-time profession, young ETA members serve an average of only three years before they are rotated back into the mainstream of society.

Russell and Miller found that more than two-thirds of the terrorists surveyed came from middle-class or even upper-class backgrounds. With the main exception of large guerrilla/terrorist organizations such as the FARC, the PKK, the LTTE, and the Palestinian or Islamic fundamentalist terrorist organizations, terrorists come from middle-class families. European and Japanese terrorists are more likely the products of affluence and higher education than of poverty.

For example, the RAF and Red Brigades were composed almost exclusively of middle-class dropouts, and most JRA members were from middle-class families and were university dropouts. Well-off young people, particularly in the United States, West Europe, and Japan, have been attracted to political radicalism out of a profound sense of guilt over the plight of the world's largely poor population. The backgrounds of the Baader-Meinhof Gang's members illustrate this in particular: Suzanne Albrecht, daughter of a wealthy maritime lawyer; Baader, the son of an historian; Meinhof, the daughter of an art historian; Horst Mahler, the son of a dentist; Holger Meins, the son of a business executive. According to Russell and Miller, about 80 percent of the Baader-Meinhof Gang had university experience.

Major exceptions to the middle- and upper-class origins of terrorist groups in general include three large organizations examined in this study—the FARC, the LTTE, and the PKK—as well as the paramilitary groups in Northern Ireland. Both the memberships of the Protestant groups, such as the Ulster Volunteer Force, and the Catholic groups, such as the Official IRA, the Provisional IRA, and the Irish National Liberation Army (INLA), are almost all drawn from the working class. These paramilitary groups are also different in that their members normally do not have any university education. Although Latin America has been an exception, terrorists in much of the developing world tend to be drawn from the lower sections of society. The rank and file of Arab terrorist organizations include substantial numbers of poor people, many of them homeless refugees. Arab terrorist leaders are almost all from the middle and upper classes.

Terrorists are generally people who feel alienated from society and have a grievance or regard themselves as victims of an injustice. Many are dropouts. They are devoted to their political or religious cause and do not regard their violent actions as criminal. They are loyal to each other but will deal with a disloyal member more harshly than with the enemy. They are people with cunning, skill, and initiative, as well as ruthlessness. In order to be initiated into the group, the new recruit may be expected to perform an armed robbery or murder. They show no fear, pity, or remorse. The sophistication of the terrorist will vary depending on the significance and context of the terrorist action.

The Colombian hostage-takers who infiltrated an embassy party and the Palace of Justice, for example, were far more sophisticated than would be, for example, Punjab terrorists who gun down bus passengers. Terrorists have the ability to use a variety of weapons, vehicles, and communications equipment and are familiar with their physical environment, whether it be a 747 jumbo jet or a national courthouse. A terrorist will rarely operate by himself/herself or in large groups, unless the operation requires taking over a large building, for example.

Members of Right-wing terrorist groups in France and Germany, as elsewhere, generally tend to be young, relatively uneducated members of the lower classes. Ferracuti and F. Bruno (1981:209) list nine psychological traits common to right-wing terrorists: ambivalence toward authority; poor and defective insight; adherence to conventional behavioral patterns; emotional detachment from the consequences of their actions; disturbances in sexual identity with role uncertainties; superstition, magic, and stereotyped thinking; metro- and auto-destructiveness; low-level educational reference patterns; and perception of weapons as fetishes and adherence to violent sub cultural norms. These traits make up what Ferracuti and Bruno call an "authoritarian-extremist personality." They conclude that right-wing terrorism may be more dangerous than left-wing terrorism because "in right-wing terrorism, the individuals are frequently psychopathological and the ideology is empty: ideology is outside reality, and the terrorists are both more normal and more fanatical."

In the past, most terrorists have been unmarried. Russell and Miller found that, according to arrest statistics, more than 75 to 80 percent of terrorists in the various regions in the late 1970s were single. Encumbering family responsibilities are generally precluded by requirements for mobility, flexibility, initiative, security, and total dedication to a revolutionary cause. Roughly 20 percent of foreign terrorist group memberships apparently consisted of married couples, if Russell and Miller's figure on single terrorists was accurate.

Terrorists are healthy and strong but generally undistinguished in appearance and manner. The physical fitness of some may be enhanced by having had extensive commando training. They tend to be of medium height and build to blend easily into crowds. They tend not to have abnormal physiognomy and peculiar features, genetic or acquired, that would facilitate their identification. Their dress and hair styles are inconspicuous. In addition to their normal appearance, they talk and behave like normal people. They may even be well dressed if, for example, they need to be in the first-class section of an airliner targeted for hijacking. They may resort to disguise or plastic surgery depending on whether they are on police wanted posters.

If a terrorist's face is not known, it is doubtful that a suspected terrorist can be singled out of a crowd only on the basis of physical features. Unlike the *yakuza* (mobsters) in Japan, terrorists generally do not have distinguishing physical features such as colorful tatoos. For example, author Christopher Dobson (1975) describes the Black September's Salah Khalef ("Abu Iyad") as "of medium height and sturdy build, undistinguished in a crowd." When Dobson, hoping for an interview, was introduced to him in Cairo in the early 1970s Abu Iyad made "so little an impression" during the brief encounter that Dobson did not realize until later that he had already met Israel's most-wanted terrorist. Another example is Imad Mughniyah, head of Hizballah's special operations, who is described by Hala Jaber (1997:120), as "someone you would pass in the street without even noticing or giving a second glance."

Guerrilla/terrorist organizations have tended to recruit members from the areas where they are expected to operate because knowing the area of operation is a basic principle of urban terrorism and guerrilla warfare. According to Russell and Miller, about 90 percent of the Argentine ERP and Montoneros came from the Greater Buenos Aires area. Most of Marighella's followers came from Recife, Rio de Janeiro, Santos, and São Paulo. More than 70 percent of the Tupamaros were natives of Montevideo. Most German and Italian terrorists were from urban areas: the Germans from Hamburg and West Berlin; the Italians from Genoa, Milan, and Rome.

Most terrorists are male. Well over 80 percent of terrorist operations in the 1966-76 period were directed, led, and executed by males. The number of arrested female terrorists in Latin America suggested that female membership was less than 16 percent. The role of women in Latin American groups such as the Tupamaros was limited to intelligence collection, serving as couriers or nurses, maintaining safe houses, and so forth.

Various terrorism specialists have noted that the number of women involved in terrorism has greatly exceeded the number of women involved in crime. However, no statistics have been offered to substantiate this assertion. Considering that the number of terrorist actions perpetrated worldwide in any given year is probably minuscule in comparison with the common crimes committed in the same period, it is not clear if the assertion is correct. Nevertheless, it indeed seems as if more women are involved in terrorism than actually are, perhaps because they tend to get more attention than women involved in common crime.

Although Russell and Miller's profile is more of a sociological than a psychological profile, some of their conclusions raise psychological issues, such as why women played a more prominent role in left-wing terrorism in the 1966-76 period than in violent crime in general. Russell and Miller's data suggest that the terrorists examined were largely males, but the authors also note the secondary support role played by women in most terrorist organizations, particularly the Uruguayan Tupamaros and several European groups. For example, they point out that women constituted one-third of the personnel of the RAF and June Second Movement, and that nearly 60 percent of the RAF and June Second Movement who were at large in August 1976 were women.

Russell and Miller's contention that "urban terrorism remains a predominantly male phenomenon," with women functioning mainly in a secondary support role, may underestimate the active, operational role played by women in Latin American and West European terrorist organizations in the 1970s and 1980s. Insurgent groups in Latin America in the 1970s and 1980s reportedly included large percentages of female combatants: 30 percent of the Sandinista National Liberation Front (FSLN) combatants in Nicaragua by the late 1970s; one-third of the combined forces of the Farabundo Martí National Liberation Front (FMLN) in El Salvador; and one-half of the Shining Path terrorists in Peru. However, because these percentages may have been inflated by the insurgent groups to impress foreign feminist sympathizers, no firm conclusions can be drawn in the absence of reliable statistical data.

Nevertheless, women have played prominent roles in numerous urban terrorist operations in Latin America. For example, the second in command of the Sandinista takeover of Nicaragua's National Palace in Managua, Nicaragua, in late August 1979 was Dora María Téllez Argüello. Several female terrorists participated in the takeover of the Dominican Embassy in Bogotá, Colombia, by the 19th of April Movement (M-19) in 1980, and one of them played a major role in the hostage negotiations. The late Mélida Anaya Montes ("Ana María") served as second in command of the People's Liberation Forces (Fuerzas Populares de Liberación—FPL) prior to her murder at age 54 by FPL rivals in 1983. Half of the 35 M-19 terrorists who raided Colombia's Palace of Justice on November 6, 1985, were women, and they were among the fiercest fighters.

Leftist terrorist groups or operations in general have frequently been led by women. Many women joined German terrorist groups. Germany's Red Zora, a terrorist group active between the late 1970s and 1987, recruited only women and perpetrated many terrorist actions. In 1985 the RAF's 22 core activists included 13 women. In 1991 women formed about 50 percent of the RAF membership and about 80 percent of the group's supporters, according to MacDonald. Of the eight individuals on Germany's "Wanted Terrorists" list in 1991, five were women.

Of the 22 terrorists being hunted by German police that year, 13 were women. Infamous German female terrorist leaders have included Susanne Albrecht, Gudrun Ensselin\Esslin, and Ulrike Meinhof of the Baader-Meinhof Gang. There are various theories as to why German women have been so drawn to violent groups. One is that they are more emancipated and liberated than women in other European countries. Another, as suggested to Eileen MacDonald by Astrid Proll, an early member of the Baader-Meinhof Gang, is that the anger of German women is part of a national guilt complex, the feeling that if their mothers had had a voice in Hitler's time many of Hitler's atrocities would not have happened.

Other noted foreign female terrorists have included Fusako Shigenobu of the JRA (Shigenobu, 53, was reported in April 1997 to be with 14 other JRA members—two other women and 12 men—training FARC guerrillas in terror tactics in the Urabá Region of Colombia); Norma Ester Arostito, who cofounded the Argentine Montoneros and served as its chief ideologist until her violent death in 1976; Margherita Cagol and Susana Ronconi of the Red Brigades; Ellen Mary Margaret McKearney of the IRA; Norma Ester Arostito of the Montoneros; and Geneveve Forest Tarat of the ETA, who played a key role in the spectacular ETA-V bomb assassination of Premier Admiral Carrero Blanco on December 20, 1973, as well as in the bombing of the Café Rolando in Madrid in which 11 people were killed and more than 70 wounded on September 13, 1974. ETA members told journalist Eileen MacDonald that ETA has always had female commandos and operators. Women make up about 10 percent of imprisoned ETA members, so that may be roughly the percentage of women in ETA ranks.

Infamous female commandos have included Leila Khaled, a beautiful PFLP commando who hijacked a TWA passenger plane on August 29, 1969, and then blew it up after evacuating the passengers, without causing any casualties. One of the first female terrorists of modern international terrorism, she probably inspired hundreds of other angry young women around the world who admired the thrilling pictures of her in newspapers and magazines worldwide showing her cradling a weapon, with her head demurely covered. Another PFLP female hijacker, reportedly a Christian Iraqi, was sipping champagne in the cocktail bar of a Japan Air Lines Jumbo jet on July 20, 1973, when the grenade that she was carrying strapped to her waist exploded, killing her.

Women have also played a significant role in Italian terrorist groups. Leonard Weinberg and William Lee Eubank (1987: 248-53) have been able to quantify that role by developing a data file containing information on about 2,512 individuals who were arrested or wanted by police for terrorism from January 1970 through June 1984. Of those people, 451, or 18 percent, were female. Of those females, fewer than 10 percent were affiliated with neo fascist groups. The rest belonged to leftist terrorist groups, particularly the Red Brigades, which had 215 female members.

Weinberg and Eubank found that the Italian women surveyed were represented at all levels of terrorist groups: 33 (7 percent) played leadership roles and 298 (66 percent) were active "regulars" who took part in terrorist actions. Weinberg and Eubank found that before the women became involved in terrorism they tended to move from small and medium-sized communities to big cities. The largest group of the women (35 percent) had been students before becoming terrorists, 20 percent had been teachers, and 23 percent had held white-collar jobs as clerks, secretaries, technicians, and nurses . Only a few of the women belonged to political parties or trade union organizations, whereas 80 (17 percent) belonged to leftist extra parliamentary movements. Also noteworthy is the fact that 121 (27 percent) were related by family to other terrorists. These researchers concluded that for many women joining a terrorist group resulted from a small group or family decision.

German intelligence officials told Eileen MacDonald that "absolute practicality...was particularly noticeable with women revolutionaries." By this apparently was meant coolness under pressure. However, Germany's female terrorists, such as those in the Baader-Meinhof Gang, have been described by a former member as "all pretty male-dominated; I mean they had male characteristics." These included interests in technical things, such as repairing cars, driving, accounting, and organizing. For example, the RAF's Astrid Proll was a first-rate mechanic, Gudrun Ensslin was in charge of the RAF's finances, and Ulrike Meinhof sought out apartments for the group.

According to Christian Lochte, the Hamburg director of the Office for the Protection of the Constitution, the most important qualities that a female member could bring to terrorist groups, which are fairly unstable, were practicality and pragmatism: "In wartime women are much more capable of keeping things together," Lochte told MacDonald. "This is very important for a group of terrorists, for their dynamics. Especially a group like the RAF, where there are a lot of quarrels about strategy, about daily life. Women come to the forefront in such a group, because they are practical."

Galvin points out the tactical value of women in a terrorist group. An attack by a female terrorist is normally less expected than one by a man. "A woman, trading on the impression of being a mother, nonviolent, fragile, even victim like, can more easily pass scrutiny by security forces...." There are numerous examples illustrating the tactical surprise factor that can be achieved by female terrorists. A LTTE female suicide commando was able to get close enough to Indian Prime Minister Rajiv Gandhi on May 21, 1991, to garland him with flowers and then set off her body bomb, killing him, herself, and 17 others.

Nobody suspected the attractive Miss Kim of carrying a bomb aboard a Korean Air Flight 858. And Leila Khaled, dressed in elegant clothes and strapped with grenades, was able to pass through various El Al security checks without arousing suspicion. Female terrorists have also been used to draw male targets into a situation in which they could be kidnapped or assassinated.

Lochte also considered female terrorists to be stronger, more dedicated, faster, and more ruthless than male terrorists, as well as more capable of withstanding suffering because "They have better nerves than men, and they can be both passive and active at the same time." The head of the German counterterrorist squad told MacDonald that the difference between the RAF men and women who had been caught after the fall of the Berlin Wall was that the women had been far more reticent about giving information than the men, and when the women did talk it was for reasons of guilt as opposed to getting a reduced prison sentence, as in the case of their male comrades.

According to MacDonald, since the late 1960s, when women began replacing imprisoned or interned male IRA members as active participants, IRA women have played an increasingly important role in "frontline" actions against British troops and Protestant paramilitary units, as well as in terrorist actions against the British public. As a result, in the late 1960s the IRA merged its separate women's sections within the movement into one IRA. MacDonald cites several notorious IRA women terrorists. They include Marion Price, 19, and her sister (dubbed "the Sisters of Death"), who were part of the IRA's 1973 bombing campaign in London. In the early 1970s, Dr. Rose Dugdale, daughter of a wealthy English family, hijacked a helicopter and used it to try to bomb a police barracks.

In 1983 Anna Moore was sentenced to life imprisonment for her role in bombing a Northern Ireland pub in which 17 were killed. Ella O'Dwyer and Martina Anderson, 23, a former local beauty queen, received life sentences in 1986 for their part in the plot to bomb London and 16 seaside resorts. Another such terrorist was Mairead Farrell, who was shot dead by the SAS in Gibraltar in 1988. A year before her death, Farrell, who was known for her strong feminist views, said in an interview that she was attracted to the IRA because she was treated the same as "the lads." As of 1992, Evelyn Glenholmes was a fugitive for her role in a series of London bombings.

MacDonald interviewed a few of these and a number of other female IRA terrorists, whom she described as all ordinary, some more friendly than others. Most were unmarried teenagers or in their early twenties when they became involved in IRA terrorism. None had been recruited by a boyfriend. When asked why they joined, all responded with "How could we not?" replies. They all shared a hatred for the British troops (particularly their foul language and manners) and a total conviction that violence was justified. One female IRA volunteer told MacDonald that "Everyone is treated the same. During training, men and women are equally taught the use of explosives and weapons."

Female terrorists can be far more dangerous than male terrorists because of their ability to focus single-mindedly on the cause and the goal. Lochte noted that the case of Susanne Albrecht demonstrated this total dedication to a cause, to the exclusion of all else, even family ties and upbringing. The RAF's Suzanne Albrecht, daughter of a wealthy maritime lawyer, set up a close family friend, Jurgen Ponto, one of West Germany's richest and most powerful men and chairman of the Dresden Bank, for assassination in his home, even though she later admitted to having experienced nothing but kindness and generosity from him.

Lochte told MacDonald that if Albrecht had been a man, she would have tried to convince her RAF comrade to pick another target to kidnap. "Her attitude was," Lochte explained, "to achieve the goal, to go straight ahead without any interruptions, any faltering. This attitude is not possible with men." (Albrecht, however, reportedly was submitted to intense pressure by her comrades to exploit her relationship with the banker, and the plan was only to kidnap him rather than kill him.) After many years of observing German terrorists, Lochte concluded, in his comments to MacDonald, that women would not hesitate to shoot at once if they were cornered. "For anyone who loves his life," he told MacDonald, "it is a good idea to shoot the women terrorists first." In his view, woman terrorists feel they need to show that they can be even more ruthless than men.

Germany's neo-Nazi groups also have included female members, who have played major roles, according to MacDonald. For example, Sibylle Vorderbrügge, 26, joined a notorious neo-Nazi group in 1980 after becoming infatuated with its leader. She then became a bomb-throwing terrorist expressly to please him. According to MacDonald, she was a good example to Christian Lochte of how women become very dedicated to a cause, even more than men. "One day she had never heard of the neo-Nazis, the next she was a terrorist." Lochte commented, "One day she had no interest in the subject; the next she was 100 percent terrorist; she became a fighter overnight."

What motivates women to become terrorists? Galvin suggests that women, being more idealistic than men, may be more impelled to perpetrate terrorist activities in response to failure to achieve change or the experience of death or injury to a loved one. Galvin also argues that the female terrorist enters into terrorism with different motivations and expectations than the male terrorist. In contrast to men, who Galvin characterizes as being enticed into terrorism by the promise of "power and glory," females embark on terrorism "attracted by promises of a better life for their children and the desire to meet people's needs that are not being met by an intractable establishment."

Considering that females are less likely than males to have early experience with guns, terrorist membership is therefore a more active process for women than for men because women have more to learn. In the view of Susana Ronconi, one of Italy's most notorious and violent terrorists in the 1970s, the ability to commit violence did not have anything to do with gender. Rather, one's personality, background, and experience were far more important. Companionship is another motivating factor in a woman's joining a terrorist group. MacDonald points out that both Susanna Ronconi and Ulrike Meinhof "craved love, comradeship, and emotional support" from their comrades.

Feminism has also been a motivating ideology for many female terrorists. Many of them have come from societies in which women are repressed, such as Middle Eastern countries and North Korea, or Catholic countries, such as in Latin America, Spain, Ireland, and Italy. Even Germany was repressive for women when the Baader-Meinhof Gang emerged.

In profiling the terrorist, some generalizations can be made on the basis on this examination of the literature on the psychology and sociology of terrorism published over the past three decades. One finding is that, unfortunately for profiling purposes, there does not appear to be a single terrorist personality . This seems to be the consensus among terrorism psychologists as well as political scientists and sociologists. The personalities of terrorists may be as diverse as the personalities of people in any lawful profession. There do not appear to be any visibly detectable personality traits that would allow authorities to identify a terrorist.

Another finding is that the terrorist is not diagnosable psychopathic or mentally sick. Contrary to the stereotype that the terrorist is a psychopath or otherwise mentally disturbed, the terrorist is actually quite sane, although deluded by an ideological or religious way of viewing the world. The only notable exceptions encountered in this study were the German anarchist terrorists, such as the Baader-Meinhof Gang and their affiliated groups. The German terrorists seem to be a special case, however, because of their inability to come to terms psychologically and emotionally with the shame of having parents who were either passive or active supporters of Hitler.

The highly selective terrorist recruitment process explains why most terrorist groups have only a few pathological members. Candidates who exhibit signs of psychopathy or other mental illness are deselected in the interest of group survival. Terrorist groups need members whose behavior appears to be normal and who would not arouse suspicion. A member who exhibits traits of psychopathy or any noticeable degree of mental illness would only be a liability for the group, whatever his or her skills. That individual could not be depended on to carry out the assigned mission. On the contrary, such an individual would be more likely to sabotage the group by, for example, botching an operation or revealing group secrets if captured. Nor would a psychotic member be likely to enhance group solidarity. A former PKK spokesman has even stated publicly that the PKK's policy was to exclude psychopaths.

This is not to deny, however, that certain psychological types of people may be attracted to terrorism. In his examination of autobiographies, court records, and rare interviews, Jerrold M. Post (1990:27) found that "people with particular personality traits and tendencies are drawn disproportionately to terrorist careers." Authors such as Walter Laqueur, Post notes, "have characterized terrorists as action-oriented, aggressive people who are stimulus-hungry and seek excitement." Even if Post and some other psychologists are correct that individuals with narcissistic personalities and low self-esteem are attracted to terrorism, the early psychological development of individuals in their pre-terrorist lives does not necessarily mean that terrorists are mentally disturbed and can be identified by any particular traits associated with their early psychological backgrounds.

Many people in other high-risk professions, including law enforcement, could also be described as "action-oriented, aggressive people who are stimulus-hungry and seek excitement." Post's views notwithstanding, there is actually substantial evidence that terrorists are quite sane. Although terrorist groups are highly selective in whom they recruit, it is not inconceivable that a psychopathic individual can be a top leader or the top leader of the terrorist group. In fact, the actions and behavior of the ANO's Abu Nidal, the PKK's Abdullah Ocalan, the LTTE's Velupillai Prabhakaran, the FARC's Jorge Briceño Suárez, and Aum Shinrikyo's Shoko Asahara might lead some to believe that they all share psychopathic or sociopathic symptoms.

Nevertheless, the question of whether any or all of these guerrilla/terrorist leaders are psychopathic or sociopathic is best left for a qualified psychologist to determine. If the founder of a terrorist group or cult is a psychopath, there is little that the membership could do to remove him, without suffering retaliation. Thus, that leader may never have to be subjected to the group's standards of membership or leadership. In addition to having normal personalities and not being diagnosable mentally disturbed, a terrorist's other characteristics make him or her practically indistinguishable from normal people, at least in terms of outward appearance. Terrorist groups recruit members who have a normal or average physical appearance.

As a result, the terrorist's physical appearance is unlikely to betray his or her identity as a terrorist, except in cases where the terrorist is well known, or security personnel already have a physical description or photo. A terrorist's physical features and dress naturally will vary depending on race, culture, and nationality. Both sexes are involved in a variety of roles, but men predominate in leadership roles. Terrorists tend to be in their twenties and to be healthy and strong; there are relatively few older terrorists, in part because terrorism is a physically demanding occupation. Training alone requires considerable physical fitness. Terrorist leaders are older, ranging from being in their thirties to their sixties.

The younger terrorist who hijacks a jetliner, infiltrates a government building, lobs a grenade into a sidewalk café, attempts to assassinate a head of state, or detonates a body-bomb on a bus will likely be appropriately dressed and acting normal before initiating the attack. The terrorist needs to be inconspicuous in order to approach the target and then to escape after carrying out the attack, if escape is part of the plan.

The suicide terrorist also needs to approach a target inconspicuously. This need to appear like a normal citizen would also apply to the FARC, the LTTE, the PKK, and other guerrilla organizations, whenever they use commandos to carry out urban terrorist operations. It should be noted that regular FARC, LTTE, and PKK members wear uniforms and operate in rural areas. These three groups do, however, also engage in occasional acts of urban terrorism, the LTTE more than the FARC and PKK. On those occasions, the LTTE and PKK terrorists wear civilian clothes. FARC guerrillas are more likely to wear uniforms when carrying out their acts of terrorism, such as kidnappings and murders, in small towns.

Terrorist and guerrilla groups do not seem to be identified by any particular social background or educational level. They range from the highly educated and literate intellectuals of the 17 November Revolutionary Organization (17N) to the scientifically savvy "ministers" of the Aum Shinrikyo terrorist cult, to the peasant boys and girls forcibly inducted into the FARC, the LTTE, and the PKK guerrilla organizations. Most terrorist leaders have tended to be well educated. Examples include Illich Ramírez Sánchez ("The Jackal") and the Shining Path's Abimael Guzmán Reynoso, both of whom are currently in prison. Indeed, terrorists are increasingly well educated and capable of sophisticated, albeit highly biased, political analysis.

In contrast to Abu Nidal, for example, who is a relatively uneducated leader of the old generation and one who appears to be motivated more by vengefulness and greed than any ideology, the new generation of Islamic terrorists, be they key operatives such as the imprisoned Ramzi Yousef, or leaders such as Osama bin Laden, are well educated and motivated by their religious ideologies. The religiously motivated terrorists are more dangerous than the politically motivated terrorists because they are the ones most likely to develop and use weapons of weapons of mass destruction (WMD) in pursuit of their messianic or apocalyptic visions. The level of intelligence of a terrorist group's leaders may determine the longevity of the group. The fact that the 17 November group has operated successfully for a quarter century must be indicative of the intelligence of its leaders.

In short, a terrorist will look, dress, and behave like a normal person, such as a university student, until he or she executes the assigned mission. Therefore, considering that this physical and behavioral description of the terrorist could describe almost any normal young person, terrorist profiling based on personality, physical, or sociological traits would not appear to be particularly useful. If terrorists cannot be detected by personality or physical traits, are there other early warning indicators that could alert security personnel? The most important indicator would be having intelligence information on the individual, such as a "watch list," a description, and a photo, or at least a threat made by a terrorist group. Even a watch-list is not fool-proof, however, as demonstrated by the case of Sheikh Omar Abdel Rahman, who, despite having peculiar features and despite being on a terrorist watch-list, passed through U.S. Customs unhindered.

Unanticipated stress and nervousness may be a hazard of the profession, and a terrorist's nervousness could alert security personnel in instances where, for example, a hijacker is boarding an aircraft, or hostage-takers posing as visitors are infiltrating a government building. The terrorist undoubtedly has higher levels of stress than most people in lawful professions. However, most terrorists are trained to cope with nervousness. Female terrorists are known to be particularly cool under pressure. Leila Khaled and Kim Hyun Hee mention in their autobiographies how they kept their nervousness under control by reminding themselves of, and being totally convinced of, the importance of their missions.

Indeed, because of their coolness under pressure, their obsessive dedication to the cause of their group, and their need to prove themselves to their male comrades, women make formidable terrorists and have proven to be more dangerous than male terrorists. Hizballah, the LTTE, and PKK are among the groups that have used attractive young women as suicide body-bombers to great effect. Suicide body-bombers are trained to be totally at ease and confident when approaching their target, although not all suicide terrorists are able to act normally in approaching their target.

International terrorists generally appear to be predominately either leftist or Islamic. A profiling system could possibly narrow the statistical probability that an unknown individual boarding an airliner might be a terrorist if it could be accurately determined that most terrorists are of a certain race, culture, religion, or nationality. In the absence of statistical data, however, it cannot be determined here whether members of any particular race, religion, or nationality are responsible for most acts of international terrorism. Until those figures become available, smaller-scale terrorist group profiles might be more useful.

For example, a case could be made that U.S. Customs personnel should give extra scrutiny to the passports of young foreigners claiming to be "students" and meeting the following general description: physically fit males in their early twenties of Egyptian, Jordanian, Yemeni, Iraqi, Algerian, Syrian, or Sudanese nationality, or Arabs bearing valid British passports, in that order. These characteristics generally describe the core membership of Osama bin Laden's Arab "Afghans" (see Glossary), also known as the Armed Islamic Movement (AIM), who are being trained to attack the United States with WMD.

This review of the academic literature on terrorism suggests that the psychological approach by itself is insufficient in understanding what motivates terrorists, and that an interdisciplinary approach is needed to more adequately understand terrorist motivation. Terrorists are motivated not only by psychological factors but also very real political, social, religious, and economic factors, among others. These factors vary widely. Accordingly, the motivations, goals, and ideologies of ethnic separatist, anarchist, social revolutionary, religious fundamentalist, and new religious terrorist groups differ significantly. Therefore, each terrorist group must be examined within its own cultural, economic, political, and social context in order to better understand the motivations of its individual members and leaders and their particular ideologies.

Although it may not be possible to isolate a so-called terrorist personality, each terrorist group has its own distinctive mindset. The mindset of a terrorist group reflects the personality and ideology of its top leader and other circumstantial traits, such as typology (religious, social revolutionary, separatist, anarchist, and so forth), a particular ideology or religion, culture, and nationality, as well as group dynamics.

Jerrold Post dismisses the concept of a terrorist mindset on the basis that behavioral scientists have not succeeded in identifying it. Post confuses the issue, however, by treating the term "mindset" as a synonym for personality. The two terms are not synonymous. One's personality is a distinctive pattern of thought, emotion, and behavior that define one's way of interacting with the physical and social environment, whereas a mindset is a fixed mental attitude or a fixed state of mind.

In trying to better define mindset, the term becomes more meaningful when considered within the context of a group. The new terrorist recruit already has a personality when he or she joins the group, but the new member acquires the group's mindset only after being fully indoctrinated and familiarized with its ideology, point of view, leadership attitudes, ways of operating, and so forth. Each group will have its own distinctive mindset, which will be a reflection of the top leader's personality and ideology, as well as group type. For example, the basic mindset of a religious terrorist group, such as Hamas and Hizballah, is Islamic fundamentalism.

The basic mindset of an Irish terrorist is anti-British sectarianism and separatism. The basic mindset of an ETA member is anti-Spanish separatism. The basic mindset of a 17 November member is antiestablishment, anti-US, anti-NATO, and anti-German nationalism and Marxism-Leninism. And the basic mindset of an Aum Shinrikyo member is worship of Shoko Asahara, paranoia against the Japanese and U.S. governments, and millenarian, messianic apocalypticism. Terrorist group mindsets determine how the group and its individual members view the world and how they lash out against it. Knowing the mindset of a group enables a terrorism analyst to better determine the likely targets of the group and its likely behavior under varying circumstances. It is surprising, therefore, that the concept of the terrorist mindset has not received more attention by terrorism specialists.

It may not always be possible to profile the individual leaders of a terrorist group, as in the case of the 17 November Revolutionary Organization, but the group's mindset can be profiled if adequate information is available on the group and there is an established record of activities and pronouncements. Even though two groups may both have an Islamic fundamentalist mindset, their individual mindsets will vary because of their different circumstances. One cannot assume to have a basic understanding of the mindset of a terrorist group without having closely studied the group and its leader(s). Because terrorist groups are clandestine and shadowy, they are more difficult to analyze than guerrilla groups, which operate more openly, like paramilitary organizations.

A terrorist group is usually much smaller than a guerrilla organization, but the former may pose a more lethal potential threat to U.S. security interests than the latter by pursuing an active policy of terrorist attacks against U.S. interests. A guerrilla group such as the FARC may kidnap or kill an occasional U.S. citizen or citizens as a result of unauthorized actions by a hard-line front commander, but a terrorist group such as the 17 November Revolutionary Organization does so as a matter of policy.

Although Aum Shinrikyo, a dangerous cult, is on U.S. lists of terrorist groups and is widely feared in Japan, it still operates openly and legally, even though a number of its members have been arrested, some have received prison sentences, and others, including Shoko Asahara, have been undergoing trial. It can probably be safely assumed that

Aum Shinrikyo will resume its terrorist activities, if not in Japan then elsewhere. Indeed, it appears to be reorganizing, and whatever new form in which this hydra-headed monster emerges is not likely to be any more pleasant than its former incarnation. The question is: what is Aum Shinrikyo planning to help bring about the apocalypse that it has been predicting for the new millennium?

Knowing the mindset of a terrorist group would better enable the terrorism analyst to understand that organization's behavior patterns and the active or potential threat that it poses. Knowing the mindsets, including methods of operation, of terrorist groups would also aid in identifying what group likely perpetrated an unclaimed terrorist action and in predicting the likely actions of a particular group under various circumstances. Indeed, mindset profiling of a terrorist group is an essential mode of analysis for assessing the threat posed by the group. A terrorist group's mindset can be determined to a significant extent through a database analysis of selective features of the group and patterns in its record of terrorist attacks. A computer program could be designed to replicate the mindset of each terrorist group for this purpose.

All terrorist and guerrillas groups may be susceptible to psychological warfare aimed at dividing their political and military leaders and factions. Guerrilla organizations, however, should not be dealt with like terrorist groups. Although the FARC, the LTTE, and the PKK engage in terrorism, they are primarily guerrilla organizations, and therefore their insurgencies and accompanying terrorism are likely to continue as long as there are no political solutions. In addition to addressing the root causes of a country's terrorist and insurgency problems, effective counterterrorist and counterinsurgency strategies should seek not only to divide a terrorist or guerrilla group's political and military factions but also to reduce the group's rural bases of support through rural development programs and establishment of civil patrols in each village or town.

Another effective counterterrorist strategy would be the identification and capture of a top hard-line terrorist or guerrilla leader, especially one who exhibits psychopathic characteristics. Removing the top hard-liners of a terrorist group would allow the group to reassess the policies pursued by its captured leader and possibly move in a less violent direction, especially if a more politically astute leader assumes control. This is what appears to be happening in the case of the PKK, which has opted for making peace since the capture of its ruthless, hard-line leader, Abdullah Ocalan. A government could simultaneously help members of urban terrorist groups to defect from their groups, for example through an amnesty program, as was done so effectively in Italy.

A psychologically sophisticated policy of promoting divisions between political and military leaders as well as defections within guerrilla and terrorist groups is likely to be more effective than a simple military strategy based on the assumption that all members and leaders of the group are hard-liners. A military response to terrorism unaccompanied by political countermeasures is likely to promote cohesion within the group. The U.S. Government's focus on bin Laden as the nation's number one terrorist enemy has clearly raised his profile in the Islamic world and swelled the membership ranks of al-Qaida. Although not yet martyred, bin Laden has become the Ernesto "Che" Guevara of Islamic fundamentalism. As Post (1990:39) has explained:

When the autonomous cell comes under external threat, the external danger has the consequence of reducing internal divisiveness and uniting the group against the outside enemy....Violent societal counteractions can transform a tiny band of insignificant persons into a major opponent of society, making their "fantasy war," to use Ferracuti's apt term, a reality."

A counterterrorist policy should be tailor-made for a particular group, taking into account its historical, cultural, political, and social context, as well as the context of what is known about the psychology of the group or its leaders. The motivations of a terrorist group—both of its members and of its leaders—cannot be adequately understood outside its cultural, economic, political, and social context. Because terrorism is politically or religiously motivated, a counterterrorist policy, to be effective, should be designed to take into account political or religious factors. For example, terrorists were active in Chile during the military regime (1973-90), but counterterrorist operations by democratic governments in the 1990s have reduced them to insignificance.

The transition from military rule to democratic government in Chile proved to be the most effective counterterrorist strategy. In contrast to relatively insignificant political terrorist groups in a number of countries, Islamic terrorist groups, aided by significant worldwide support among Muslim fundamentalists, remain the most serious terrorist threat to U.S. security interests. A U.S. counterterrorist policy, therefore, should avoid making leaders like Osama bin Laden heroes or martyrs for Muslims. To that end, the eye-for-an-eye Israeli policy of striking back for each act of terrorism may be highly counterproductive when applied by the world's only superpower against Islamic terrorism, as in the form of cruise-missile attacks against, or bombings of, suspected terrorist sites.

Such actions, although politically popular at home, are seen by millions of Muslims as attacks against the Islamic religion and by people in many countries as superpower bullying and a violation of a country's sovereignty. U.S. counterterrorist military attacks against elusive terrorists may serve only to radicalize large sectors of the Muslim population and damage the U.S. image worldwide. Rather than retaliate against terrorists with bombs or cruise missiles, legal, political, diplomatic, financial, and psychological warfare measures may be more effective. Applying pressure to state sponsors may be especially effective. Cuba and Libya are two examples of terrorist state sponsors that apparently concluded that sponsoring terrorists was not in their national interests. Iran and Syria may still need to be convinced.

Jeanne Knutson was critical of the reactive and ad hóc nature of U.S. counterterrorism policy, which at that time, in the early 1980s, was considered an entirely police and security task, as opposed to "...a *politically* rational, comprehensive strategy to deal with *politically* motivated violence." She found this policy flawed because it dealt with symptoms instead of root causes and instead of eradicating the causes had increased the source of political violence. She charged that this policy routinely radicalized, splintered, and drove underground targeted U.S. groups, thereby only confirming the "we-they" split worldview of these groups. Unfortunately, too many governments still pursue purely military strategies to defeat political and religious extremist groups.

Abroad, Knutson argued, the United States joined military and political alliances to support the eradication of internal dissident groups without any clear political rationale for such a stance. She emphasized that "terrorists are individuals who commit crimes for *political* reasons," and for this reason "the political system has better means to control and eliminate their activities and even to attack their root causes than do the police and security forces working alone." Thus, she considered it politically and socially unwise to give various national security agencies, including the Federal Bureau of Investigation (FBI), the political role of choosing targets of political violence. She advocated "a necessary stance of neutrality toward national dissident causes—whether the causes involve the territory of historical friend or foe." She cited the neutral U.S. stance toward the Irish Republican Army (IRA) as a case study of how to avoid anti-U.S. terrorism. Her views still seem quite relevant.

Goals of a long-range counterterrorism policy should also include deterring alienated youth from joining a terrorist group in the first place. This may seem an impractical goal, for how does one recognize a potential terrorist, let alone deter him or her from joining a terrorist group? Actually, this is not so impractical in the cases of guerrilla organizations like the FARC, the LTTE, and the PKK, which conscript all the young people in their rural areas of operation who can be rounded up. A counter strategy could be approached within the framework of advertising and civic-action campaigns.

A U.S. government-sponsored mass media propaganda campaign undertaken in the Colombian countryside, the Kurdish enclaves, and the Vanni region of Sri Lanka and tailor-made to fit the local culture and society probably could help to discredit hard-liners in the guerrilla/terrorist groups sufficiently to have a serious negative impact on their recruitment efforts. Not only should all young people in the region be educated on the realities of guerrilla life, but a counterterrorist policy should be in place to inhibit them from joining in the first place. If they are inducted, they should be helped or encouraged to leave the group.

The effectiveness of such a campaign would depend in part on how sensitive the campaign is culturally, socially, politically, and economically. It could not succeed, however, without being supplemented by civic-action and rural security programs, especially a program to establish armed self-defense civil patrols among the peasantry. The Peruvian government was able to defeat terrorists operating in the countryside only by creating armed self-defense civil patrols that became its eyes and ears. These patrols not only provided crucial intelligence on the movements of the

Shining Path and Tupac Amaru terrorists, but also enabled the rural population to take a stand against them.

There is little evidence that direct government intervention is the major factor in the decline of terrorist groups. Clearly, it was an important factor in certain cases, such as the RAF and with various urban Marxist-Leninist group in Latin America where massive governmental repression was applied (but at unacceptably high cost in human rights abuses). Social and psychological factors may be more important. If, for security reasons, a terrorist group becomes too isolated from the population, as in the case of the RAF and the Uruguayan Tupamaros, the group is prone to losing touch with any base of support that it may have had. Without a measure of popular support, a terrorist group cannot survive.

Moreover, if it fails to recruit new members to renew itself by supporting or replacing an aging membership or members who have been killed or captured, it is likely to disintegrate. The terrorist groups that have been active for many years have a significant base of popular support. Taylor and Qualye point out that despite its atrocious terrorist violence, the Provisional IRA in 1994 continued to enjoy the electoral support of between 50,000 and 70,000 people in Northern Ireland. The FARC, the LTTE, and the PKK continue to have strong popular support within their own traditional bases of support.

In the cases of West German and Italian terrorism, counterterrorist operations undoubtedly had a significant impact on terrorist groups. Allowing terrorists an exit can weaken the group. For example, amnesty programs, such as those offered by the Italian government, can help influence terrorists to defect. Reducing support for the group on the local and national levels may also contribute to reducing the group's recruitment pool. Maxwell Taylor and Ethel Quayle have pointed out that penal policies in both countries, such as allowing convicted terrorists reduced sentences and other concessions, even including daytime furloughs from prison to hold a normal job, had a significant impact in affecting the long-term reduction in terrorist violence. Referring to Italy's 1982 Penitence Law, Taylor and Quayle explain that "This law effectively depenalized serious terrorist crime through offering incentives to terrorists to accept their defeat, admit their guilt and inform on others so that the dangers of terrorist violence could be diminished."

Similarly, Article 57 of the German Penal Code offers the possibility of reduction of sentence or suspension or deferment of sentence when convicted terrorists renounce terrorism. Former terrorists do not have to renounce their ideological convictions, only their violent methods. To be sure, these legal provisions have not appealed to hard-core terrorists, as evidenced by the apparent reactivation of the Italian Red Brigades in 1999. Nevertheless, for countries with long-running insurgencies, such as Colombia, Sri Lanka, and Turkey, amnesty programs for guerrillas are very important tools for resolving their internal wars.

With regard to guerrilla/terrorist organizations, a major question is how to encourage the political wing to constrain the military wing, or how to discredit or neutralize the military branch. The PKK should serve as an ongoing case study in this regard. Turkey, by its policy of demonizing the PKK and repressing the Kurdish population in its efforts to combat it instead of seeking a political solution, only raised the PKK's status in the eyes of the public and lost the hearts and minds of its Kurdish population. Nevertheless, by capturing Ocalan and by refraining thus far from making him a martyr by hanging him, the Turkish government has inadvertently allowed the PKK to move in a more political direction as advocated by its political leaders, who now have a greater voice in decision-making.

Thus, the PKK has retreated from Turkey and indicated an interest in pursuing a political as opposed to a military strategy. This is how a guerrilla/terrorist organization should end, by becoming a political party, just as the M-19 did in Colombia and the Armed Forces of National Liberation (FALN) did in El Salvador.

2

News and Terrorism

As a Critical Infrastructure

Since the terrorist attacks of September 11, 2001, the U.S. government has repeatedly warned its citizens that a similar (or even more deadly) attack within its borders is not a matter of if, but when. Far-reaching efforts have been made to prevent and prepare for such a crisis. Officials are working to harden every conceivable critical infrastructure target, but at first, one was overlooked: the news media. Maybe that is because few people think of the news media as a part of their nation's critical infrastructure, and in the United States, it is (thankfully) outside government regulation.

When we think of infrastructure, we usually think of tangible things like energy pipelines, transportation systems, and computer networks. With the advent of modern terrorism, the news media also belong in this category. They are the main communication conduit to any nation's most important infrastructure: its citizens. While the connectedness of modern infrastructures means greater efficiency, it also creates new types of interdependencies that expose new vulnerabilities. The news media may in fact be the weakest link in the system. We need to protect the media as zealously as we protect the electric power grid and nuclear reactors, and not just their printing plants and broadcast towers.

The agencies and officials working to bolster homeland defense need to work more closely with journalists and with organizations like the U.S. National Academies. Journalists need to be armed with the knowledge to work effectively as part of the nation's response to terrorism. To do that, they need the help of the engineering, science, and medical communities. At the National Academy of Engineering (NAE), we wrestled with the question of how to help the media become better informed and more conscious of their importance during a terrorist attack. Journalists in the United States are constitutionally protected and vigorously independent. No one can dictate what stories they choose or how they are reported.

Many in the U.S. government (probably in all governments) think of journalists as pests, even as threats to national security. The feeling is often that reporters should be avoided as much as possible and told as little as possible. Especially in current times, the opposite is true. A study by the New York Academy of Medicine says that "far fewer people than needed would follow protective instructions" during terrorist attacks involving smallpox or a radiological bomb. People will not blindly do as the government tells them; they need to understand the reasons for actions being taken. In the midst of a terrorist attack involving weapons of mass destruction, effectively communicating potentially complex information will be a difficult challenge that will fall largely upon the news media.

Getting good information to the public in the midst of a crisis can actually be more vital than the actions of first responders. In fact, journalists are first responders. Not only do they sometimes arrive at the scene first but they are the only ones focused on and able to communicate risk to people in real time. They can save lives through efficient delivery of accurate information. Yet, with today's competitive 24-hour news coverage, journalists are under tremendous pressure to fill airtime and print space and to get the story first. Of course, this can lead to speculation, and it is not always harmless. Sometimes it can cost lives. This is not just the media's problem; it is not just the government's problem. It is the engineering and science communities' problem, too.

The NAE decided to conduct a tabletop terrorism scenario exercise that would, for the first time, focus on communication issues. The goals would include simply bringing together groups that do not often work together—journalists, scientists, government officials—to meet and begin to understand each other's needs during the chaos of a terrorist attack. Situations can look much different when viewed from another perspective, and the time to start a relationship is not during a crisis.

Government officials must understand the pressures journalists face when trying to report relevant information under the pressure of continuous coverage during a crisis. Journalists, in turn, must better realize the need of government spokespeople to be cautious. Scientists should be prepared to deviate from their ultra precise tendencies and convey their expertise in ways that laypeople understand. Journalists, in particular, have few precedents for this new type of warfare— it is different from traditional war reporting. They need a strategy to deal with it and an instant pool of trusted experts who are good communicators.

On June 20, 2003, the National Academies hosted, in association with the Greater Washington, D.C., Board of Trade and the U.S. Department of Homeland Security (DHS), a terrorism scenario simulation, entitled Media and the First Response, which engaged all of the above groups. The event was so successful that the DHS provided funding for the National Academies to conduct similar workshops in 10 cities across the United States. The

National Academies worked closely with both DHS and the Radio-Television News Directors Foundation (RTNDF) to create this series, called News and Terrorism: Communicating in a Crisis.

Helping the Pieces Work Together

The nationwide News and Terrorism: Communicating in a Crisis workshops began in August 2004. Approximately 100 invitees attend in each city, mainly journalists; government officials; science, engineering, and medical experts; and emergency responders. The workshops include a background presentation on the science and technology behind potential terrorism threats and information on self-protection when near various terrorist incidents. The centerpiece is an exercise in which seven to eight people (journalists, government officials, and a technical expert) react to an unfolding terrorism scenario.

Scenario simulations are powerful tools for vividly illustrating problems and uncovering system weaknesses. Commonly used by the military for many years, such exercises are being used more and more frequently by emergency officials and even private industries. It is difficult to prepare for events that have not happened before. Thinking through the communication flow in a crisis, before it actually occurs, is vital in this information age. The public expects to be informed right away, and they will be. The main questions are: by whom, and how well?

Examples of dialog from our scenario exercises cannot be shared, because the ground rules require that the comments of participants not be for attribution in any public forum. This allows the players in our war games to be open and honest, knowing that their candid remarks will not appear in the newspaper the next day. While much of the dialog might actually increase public confidence in the preparedness of those tasked with keeping us informed, too much of it would certainly cause unease.

A demonstration of the kind of interactions revealed during these workshops need not come from the exercises themselves, however. Real life provides very good examples. On September 30, 2001, U.S. Health and Human Services Secretary Tommy Thompson appeared on the CBS television network program *60 Minutes* and said, "We're prepared to take care of any contingency, any consequence that develops for any kind of bioterrorism attack." He did not know that such an attack was already under way, and his remarks would soon be severely tested.

On October 4, 2001, Florida officials announced that tabloid magazine editor Bob Stevens had been diagnosed with inhalation anthrax. Later that day, Secretary Thompson made a rare appearance at a White House press briefing. He was joined by an expert in biological threats, Dr. Scott Lillibridge.

After making a brief statement about the situation involving Mr. Stevens, Secretary Thompson was peppered with questions from journalists. The second one asked whether Mr. Stevens had come in contact with raw wool or if he might be a gardener, to which Secretary Thompson responded:

SECRETARY THOMPSON: We have the FBI [Federal Bureau of Investigation] and we have dispatched, … as soon as we heard anything suspicious, we have our CDC [Centers for Disease Control and Prevention] officials there, on the ground. And they are going to go through—the last couple weeks, go to the restaurants. He traveled to North Carolina. We've also dispatched people from CDC to North Carolina, to the communities that he was there. We're checking with his neighbors. We're investigating with the FBI all known places and all the things that he might have ingested.

JOURNALIST: Mr. Secretary, what are some of the sources that could cause such an infection?

SECRETARY THOMPSON: That's why the doctor is here. And do you want to answer that?

DR. LILLIBRIDGE: Sure. Sporadic cases may occur from contact with wool, animal products, hides, that sort of thing. And occasionally we don't know the context of these. These are sporadic, episodic things that happen from time to time.

JOURNALIST: But how sporadic? You just named two cases last year in Texas and then Florida in 1974. That's two ...

SECRETARY THOMPSON: They're very rare. It's very rare.

JOURNALIST: So this is the third since 1974?

SECRETARY THOMPSON: We don't know that, but this is a confirmed, and at this point in time, it's an isolated case. And there is no other indications anybody else has got anthrax.

JOURNALIST: Do you know if he happened to work around wool or an, of the products that might have ...

SECRETARY THOMPSON: We don't know that at this point in time. That's entirely possible. We do know that he drank water out of a stream when he was traveling to North Carolina last week ...

Biological disease expert Dr. Lillibridge continued to be ignored during the next 13 questions, though many of them involved technical issues and questions about symptoms. Finally, Secretary Thompson turned to Dr. Lillibridge when a questioner probed about how likely it was that there had been other anthrax cases in the past year that had gone undiagnosed.

SECRETARY THOMPSON: It's entirely possible.

JOURNALIST: Possible or likely or... ?

SECRETARY THOMPSON: Would you say it's probable?

DR. LILLIBRIDGE: It's possible. As you heighten surveillance, you'll get more.

That response was the last heard from Dr. Lillibridge. The news media seemed more interested in getting a good quote from a high-profile cabinet secretary than in understanding what they need to report to the U.S. public.

It did not help that Secretary Thompson was trying his best to dismiss the possibility of terrorism while making statements uninformed by science. Either he should have been better briefed on the facts, or he was shading the truth. Either way, the public was not served well.

There were five more questions, all directed at Secretary Thompson, before White House Press Secretary Ari Fleischer wrapped up the subject of anthrax and turned the press conference to other matters, saying: "Any additional information will be made available by either the CDC or the HHS [U.S. Department of Health and Human Services]."

The White House obviously commands center stage in the middle of such crises. Are they equipped to effectively communicate the science and risks? These sorts of issues are dramatically brought to light, just as in this example, through our mock scenarios.

In the real-life case, as Mr. Fleischer tried to move on to another subject, the journalists' questions continued, asking for verification of Dr. Lillibridge's name and then following up by asking: "Is he an M.D. or a Ph.D?" Mr. Fleischer responded, "I'd have to look that up. I couldn't tell you. You may want to just check with HHS."

The scientist, who is best equipped to inform the public about potential dangers or lack thereof, was not even important enough for a White House follow-up. The journalists, too, missed many opportunities to get solid facts from him. These are lessons that should have been learned in post-analysis of the 2001 anthrax attacks, but that is not yet evident. Similar mistakes during our scenario-based workshops are sometimes frightening. Hopefully those in attendance are now doing something about it.

Let us return to the real-life events. On October 15, 2001, less than two weeks after the Stevens diagnosis and just a few days after it was announced that an NBC News employee in New York had also been infected with anthrax, an intern in U.S. Senator Thomas Daschle's office opened an envelope containing white powder that tests confirm contained anthrax. The next day, Senator Daschle stepped before the media's microphones and began talking about a foreign operations bill. Of course, the first question from journalists was about anthrax:

JOURNALIST: Senator, there are a dozen Senate offices closed today. There is no mail. People are being tested for anthrax. How is the Senate functioning?

SENATOR DASCHLE: I think the Senate is functioning as we could hope it would. Obviously, these are difficult times, and we are going to have logistic and administrative challenges that we're going to have to face. But I think understand [*sic*] the circumstances, the Senate is functioning quite well. We'll be back in business in all respects within the next several days.

Although Senator Daschle lacks scientific expertise, the journalists went on to question him about technical issues:

JOURNALIST: If today it was so important to close the Hart Building partially as a precaution, why wasn't it important to do that yesterday? And do you think that people might have been put at risk by that?

SENATOR DASCHLE: I am confident that there was really no risk involved. This was simply an effort to determine whether there is even a modicum of anthrax that could be found in one of the vents or one of the air ducts that would give us some indication that there was dissemination. Keep in mind that even if there is some trace, it wouldn't be of sufficient force or strength to be of health risk to those who are exposed.

Should Senator Daschle be the spokesperson to the American public about such issues? Does he have scientific credibility? The next day, leaders of the U.S. House of Representatives held a widely broadcast news conference in which they, too, were asked about the technical nature of the anthrax.

JOURNALIST: Who briefed you on the anthrax? Who told you that it was sophisticated? Because the senators are now saying that it was garden variety anthrax.

CONGRESSMAN [J. DENNIS] HASTERT: You know, we're just saying that the way that it was distributed, with a flume, was unlike anything that we've seen up to this point. Of course, we weren't in contact or saw what happened in the building in Florida or any of those buildings in New York. But this was different from the anthrax that was just out there in an envelope on a white, powdery substance. It actually had a flume and, you know, infected a lot of people.

JOURNALIST: Is there something that propelled it?

CONGRESSMAN [RICHARD] GEPHARDT: No, it was the same as the other situations. But you had now 29 or 30 people who have tested positive. That's a new development. Obviously, that's more than we've seen in these other instances. And it led the people who have looked at this to believe that it is a higher grade, weapon-grade kind of anthrax.

Journalists too often turn toward congressional people and other high-profile celebrities with little expertise in the science at the heart of issues involving potential weapons of mass destruction. In this case, conflicting statements between public officials certainly did not reassure the public. Why were members of Congress the lead spokespeople in the midst of such a technologically complex situation? Why were not those at the forefront of the issue scientists with expertise in anthrax bacteria or airflow physics, engineers who could discuss ventilation systems, and medical professionals who could talk about patient symptoms?

Our News and Terrorism workshops are designed to help such issues bubble to the surface and create a discussion among key players about how to do it better in real life. We have also produced four-page fact sheets on radiological, chemical, biological, and nuclear threats. The purpose is mainly to help journalists (though others would find them useful, too) become knowledgeable enough not only to report accurate information to the public but also to simply ask the right questions.

Effectively Serving the Public

Most journalists are not engaged enough on the underlying issues, the substance and context behind breaking news, which often involve science and technology. So their questions are at a superficial level. The News and Terrorism scenarios are a dynamic way of pulling people into important discussions. Getting the interest of journalists when there is not an immediate threat is a challenge, but a vitally important one.

Too often the news media takes the easiest path, which often means the political angle. In part, this is because politics is a form of theater, and entertainment trumps substance in the economically driven media. Politics is also about people and personalities. The news audience, unfortunately, has been trained to have a limited and shallow attention span.

If journalists are going to report in sensational and inaccurate ways, then some might argue that journalists should simply be barred from reporting about terrorist incidents. That way, the terrorists would not have a stage. But fear of the unknown fuels terror too, and distrust in government stems from such withholding of information. People will get their news, if not from the media, then through the rumor mill. When it works correctly, independent professional journalism is the best way to inform the public. That is why the First Amendment of the U.S. Constitution is so important.

We need the media not only to become a stronger part of our infrastructure but to keep challenging the government, because that exercise makes us all stronger. However, uninformed journalists cannot effectively question authority. For example, some well-meaning government efforts to classify or withhold information could end up actually harming national security by slowing the delivery of scientific research results beneficial to society. Journalists need to be equipped to effectively question such policies, and even the work of scientists. To that end, we should help journalists become better informed.

Even without direct government interference in news reporting, new technologies are now cutting out the journalistic middleman. For example, officials can send emergency instructions directly to personal devices like cellular phones (through which person-to-person reporting, including rumors, would also be conveyed). The public should not rely on such sources alone. Unless people are well informed, through independent professional sources, they will not know how to analyze the issues and how to assess the information being provided by their leaders.

The public will not automatically follow orders from authorities. Citizens need to understand the reasons for actions they are being asked to take, and they can deal with bad news. Those who seek to calm for the sole sake of maintaining order will ultimately create the opposite effect, and the public will begin to lose trust in its government.

This is the ultimate goal of terrorists. The public must understand the truth about real dangers. People will respond well, if the conveyor of information is perceived as trustworthy. Unfortunately, right now neither the government nor journalists are held in very high regard. We must work to change that.

Firefighters, police, and government officials are not always (maybe not even usually) the most important first responders. Often that role falls on such citizens as school teachers or even the media. They all need to understand, and can handle, the truth. Authority figures should not have an information monopoly. The more people are empowered to respond appropriately, the more secure we all will be.

As a local police chief once said: "You can't build a fence around a community, but you can arm your citizens with knowledge." Scientists, engineers, and the government must work to get good information into the hands of the media quickly during a cyber-, radiological, nuclear, chemical, or biological attack. We must all work together to determine the best way of doing that.

It is at our own risk that the technical community thinks of its homeland security responsibility as simply creating the latest counterterrorism technologies. They should also help empower the media, and thus the public, with knowledge. Ignorance and misinformation can be as damaging to the information infrastructure of a nation as a break in an oil pipeline. It can cause paralysis among citizens who are often the best first responders, confuse professionals trying to respond to a crisis, and help generate the fear that is the terrorists' goal.

New Media and Military Operations

The Israeli-Hezbollah War of 2006 provides recent, glaring evidence of how the current information environment has impacted the way warfare is conducted today. Hezbollah masterfully manipulated and controlled that environment to its advantage, using (at times staged and altered) photographs and videos to garner regional and worldwide support. If this doesn't sound new, it shouldn't…especially if you are an Israeli. Hamas effectively used the same techniques to turn the Battle of Jenin in April, 2002 into not only a strategic informational victory, but a historical legend of resistance that lives on today in the hearts and minds of Palestinians. The Israelis, having won total tactical victory in Jenin, literally snatched defeat from the jaws of victory by abrogating the information battle space to Hamas.

Certainly, United States military leaders can, at a minimum, empathize with the Israelis. Insurgent use of information as an asymmetric strategic means has been extremely effective in the current theaters of Iraq and Afghanistan leading Richard Holbrooke to famously muse: "How can a man in a cave out-communicate the world's leading communications society?" Had Holbrooke even superficially studied recent history he could have answered his own question. The monopoly enjoyed by nation-states over information as an element of power was rapidly lost as technology improved and as the means to transmit that information became smaller, faster, cheaper and, consequently, ubiquitous. And the outlook in that regard certainly does not seem to favor lumbering bureaucracies any time in the future.

These enabling technological capabilities have popularly been tagged "new media". Broadly, new media has been described as "that combustible mix of 24/7 cable news, call-in radio and television programs, Internet bloggers and online websites, cell phones and iPods." But, of course this menu limits the definition to present day capabilities and is quickly outdated given current and expected future technological advances. New media in this context quickly becomes "old" media, especially in light of projected asymptotic increases in speed and capacity. So, a more timeless definition should consider new media as any capability that empowers a broad range of actors (individuals through nation-states) to create and disseminate near-real time or real time information with the ability to affect a broad (regional or worldwide) audience.

If the United States military hopes to fight and win in a future information environment dominated by new media it must fully understand both the opportunities and challenges of that environment. This includes the ability to exploit new media to achieve military objectives and defeat an adversary's skilled use of it within real and perceived bureaucratic and legal constraints. A review of these capabilities and their use reveals a requirement for a significant cultural shift within the military, while recognizing that current planning processes remain valid. It also points to the importance of competing on the information battle space, not only in counterinsurgency operations, but across the spectrum of conflict.

Today's Information Environment

The current information environment has leveled the playing field for not only nation states, but non-state actors, multinational corporations and even individuals to affect strategic outcomes with minimal information infrastructure and little capital expenditure. Anyone with a camera cell phone and personal digital device with internet capability understands this. On the other hand, the United States military has increasingly leveraged advances in information infrastructure and technology to gain advantages on the modern battlefield. One example from Operation Iraqi Freedom is the significant increase in situational awareness from network centric operations that enabled coalition forces to swiftly defeat Iraqi forces in major combat operations. Another includes the more prevalent use of visual information to record operations in order to proactively tell an accurate story or effectively refute enemy "disinformation."

Even a cursory look at advances in technology confirms what most people recognize as a result of their daily routine. The ability to access, collect, and transmit information is clearly decentralized to the lowest level (the individual). The technology is increasingly smaller, faster and cheaper. Consequently, the ability to control and verify information is much more limited than in the recent past. Nor will it get any easier.

In 1965, the physical chemist Gordon Moore, co-founder of Intel, predicted that the number of transistors on an integrated chip would double every eighteen months. Moore predicted that this trend would continue for the foreseeable future. Moore and most other experts expect Moore's Law to remain valid for at least another two decades.

So, if you're into control, as nation-states, bureaucracies and the military tend to be, the future may appear bleak since not only is the ability to access, collect and transmit information decentralized, the capacity to do the same continues to increase exponentially. With this in mind consider both the new media capabilities and methods currently used and where the future may be heading as the basis for understanding what this means for the war fighter.

The Internet

The internet is the obvious start point for any discussion of the impact of new media. It is important to note that the World Wide Web, as a subset of the internet, is essentially ungoverned, providing obvious freedoms and cautions. The web gives the individual a voice, often an anonymous voice…and a potentially vast audience. Websites are easily established, dismantled and reestablished, making them valuable to extremist movements. Islamic extremist websites grew from twenty to over 4,000 in only five years.

Web logs (blogs) are another example of the power that the internet provides to individuals along with the dilemma they pose for nation-states. There were 35.3 million blogs as of April 2006 reflecting a doubling of size every six months of the previous three years. Most of these, of course, have little effect on the conduct of nation-states or their militaries, but those that gain a following in the national security arena, can have a huge impact.

Ex President George W. Bush recently cited Iraqi bloggers to point to progress being made in Iraq, having apparently learned both the importance and value of blogs in 2004 when investigative bloggers cleared his name in the infamous CBS airing that questioned his military service. The U.S. Central Command actively engages dissident voices by participating in blogs that are critical of the war on terror noting "with the proliferation of information today, if you're not speaking to this forum, you're not being heard by it."

Video use and dissemination has skyrocketed as the capabilities of the internet have increased. The YouTube phenomenon's power and access is evidenced by its purchase for $1.6 Billion by Google only 20 months after its founding. Like blogs, YouTube serves a variety of purposes to include entertainment. But, also like blogs, YouTube can empower individuals to achieve strategic political and military effects where easy upload of their videos, without editorial oversight, allows access to a nearly unlimited audience. Thus, the use of the improvised explosive device (IED) by insurgents shifts from a military tactical weapon to a strategic information weapon when the IED detonator is accompanied by a videographer.

And again, like blogs, the United States military has recognized the importance of competing in the video medium, using YouTube to show ongoing images of U.S. operations in Iraq. While websites, bloggers and video proliferate in today's internet ("web 2.0") the "over the horizon" technology of "web 3.0" while in its infancy, is rapidly increasing in popularity. Web 3.0 is generally about being inside a 3D virtual world that is low-cost and emotive. This is the "metaverse" or virtual universe of applications like Second Life and others.

Second Life is attractive as an opportunity to socialize where there is no need to compete and can be exploited as a tool for learning. Multinational corporations see a movement where they will plan and execute business plans in the 3D internet world. But, like the other internet based applications, web 3.0 provides opportunities for darker undertakings. The virtual universes show initial signs of providing training grounds for terrorist organizations and anonymous locations for criminal money-laundering.

Mobile Technologies

The internet clearly is part of the new media phenomenon, but the internet has not penetrated large areas of the world, especially in the poorest areas of underdeveloped countries. The cell phone, however, as a means of mobile technology, is increasingly available worldwide and deserves discussion as a potentially potent capability to affect national security and military issues; arguably even more so than the internet. There are numerous examples of cell phone Short Message Service (SMS text) messaging shaping political campaigns by mobilizing and revolutionizing politics.

It is used both to call people to popular protests as well as used by governments to provide misinformation in order to quell such protests. Text messaging is the medium of choice in overseas countries. It bypasses mass media and mobilizes an already persuaded populace as a means of lightweight engagement. Cell phones currently contain the technology to text, provide news, video, sound, voice, radio and internet. Mobile is pervasive in the third world. 97% of Tanzanians have access to mobile phones. Mobile coverage exists throughout Uganda. There are 100 million handsets in sub-Saharan Africa. Radio is the only media device more prevalent than mobile.

Consider the economic implications of mobile technologies as well. 59% of mobile phones are in the developing world–over seven million mobile subscribers in Kenya alone. Efforts are under way to develop African specific mobile applications, e.g. distributing commodity prices (such as vegetable prices) to local village producers. Cell phones are used as credit cards in Kenya. You can pay for cab fare or for fish at the market with your cell phone. Cell towers are being raised in Lake Victoria to allow fisherman to call to shore with their catch numbers as they set out to market. Mobile phones are ubiquitous in Asia. There are over 400 million users in China. Farmers receive crop market prices from the Chinese government via text messaging in order to allow them to harvest at the best possible time.

Like any other new media capability cell phone technology provides opportunities and challenges. Many young Iranians are turning to cell phones as a means for political protest...an opportunity that can be exploited. On the other hand, criminals and terrorists can use cell phones to quickly organize an operation, execute it and disperse using phone cards to provide cover from being traced. On an international scale, the challenge is often in the same laws that provide individual protections in democratic societies. Witness recent court battles within the United States regarding eavesdropping on foreign conversations without a court order when those conversations may be routed through a U.S. cell phone service provider.

Mainstream Media

Mainstream media certainly takes advantage of technological advances in order to remain competitive. Marvin Kalb, in the Harvard report on the Israeli-Hezbollah War notes that:

To do their jobs, journalists employed both the camera and the computer, and, with the help of portable satellite dishes and video phones "streamed" or broadcast their reports..., as they covered the movement of troops and the rocketing of villages—often, (unintentionally, one assumes) revealing sensitive information to the enemy. Once upon a time, such information was the stuff of military intelligence acquired with considerable effort and risk; now it has become the stuff of everyday journalism. The camera and the computer have become weapons of war.

This real time reporting from the field has obvious impacts on the war fighter, but competition with new media for the first and fastest story also means that today's mainstream media is not your father's mainstream media. Because of the plethora of information available today, newspapers, which once competed for knowledge as a scarce resource, today compete for a new scarce resource: the readers' (or listeners' in the case of broadcast media) attention.

Perhaps that is why increasing numbers of young adults turn to Jon Stewart's "The Daily Show" for their news. It should come as no great shock, then, that "good news" stories about military operations do not appear with regularity in mainstream print and broadcast journalism. Good news doesn't sell...because it doesn't grab the reader's (or viewer's) attention.

Of course in an environment where the speed of breaking news means viewership and thus advertising dollars, accuracy is sometimes sacrificed as well. In a strange twist, mainstream media now turns increasingly to bloggers for their stories and the most respected bloggers require multiple sources to verify accuracy. Consequently, the distinction between new and mainstream media sources becomes blurred, leaving it to the reader, already bombarded with information, to distinguish fact from fiction (or perhaps more accurately "spin" from context).

The Impact of New Media

The impact of new media in today's information environment has significant implications for the U.S. military. The first, and perhaps most obvious, is that individuals and small groups (e.g. the terrorist franchised cell) wield information as a strategic, asymmetric weapon by very effectively leveraging new media capabilities. The lack of any bureaucratic structure further enables individual empowerment by also allowing a nimbleness of response that is the antithesis of nation-state governments. Additionally, the lack of governance of the World Wide Web allows any statements or positions to be presented without regard to truth, context or ethical foundation.

Current military doctrine espouses information superiority as the solution to the numerous dilemmas posed by new media:

To succeed, it is necessary for US forces to gain and maintain information superiority. In Department of Defense (DOD) policy, information superiority is described as the operational advantage gained by the ability to collect, process, and disseminate an uninterrupted flow of information while exploiting or denying an adversary's ability to do the same.

But this clearly is an unachievable definition of success given the challenges of the current and future information environment in which military operations take place. Individuals, empowered by new media capabilities, and unencumbered by bureaucratic processes and moral/ethical standards will continue to wield information as power asymmetrically. "Wildcards" will routinely gain resonance in the information environment and once the information is out it may be extremely difficult to refute or explain it (e.g. Abu Ghraib photos or Osama bin Laden's taped messages).

Simply consider how corrections are made in mainstream print media. The correction is typically buried deep in the publication while referring to an article that likely appeared on page one above the fold. The chances of the original reader stumbling upon the correction are obviously slim. This reflects the "genie in a bottle" syndrome of today's information environment—once information is out of the bottle trying to stuff it back in is very difficult.

The Case of Jenin

The Battle of Jenin provides a relevant case study in both the impact of "wildcards" using new media means and the strategic impact resulting from a military that does not proactively and effectively manage the information environment. Israel embarked on a military campaign in the spring of 2002 to root out entrenched Hamas terrorists in the West Bank of the Palestinian occupied territories. The Jenin operation specifically targeted approximately 200 militants operating within the city and was deemed a tactical success.

The militant's infrastructure was destroyed, and a number of insurgents were killed and arrested. At the conclusion of the campaign terrorists attacks within Israel dropped off in the near term. However, the operation proved a significant strategic failure. Reaching the outskirts of Jenin, Israeli Defense Forces (IDF) restricted the media from entering the city. Thus, the information battle space was abrogated to the militants. Framed digital photos of homes being demolished by Israeli bulldozers while frantic families looked on were transmitted quickly, and quite effectively, to the press and soon hit the international wires. Cell phones were the only means of communication to the city and so the press used uncollaborated cell phone interviews with Jenin residents to tell their stories. These interviews spoke of massacre within the city in emotive tone and content.

While later disproved, those claims caused an international withdrawal of support for Israel. Further, Jenin remains, to this day, a mythical symbol of Palestinian resistance and Israeli ruthlessness among Palestinians. The lesson of Jenin is that the military may be able to dominate the information environment in a localized geographical area for a limited period of time but these wildcards, using new media capabilities, become that limiting factor. So, if information superiority is effectively unachievable and information dominance can only be partially achieved how can the U.S. military succeed in the information battle space?

The answer is that the military can, and should be expected to *manage* the information environment on their terms. This requires playing a proactive role in shaping that environment as well as establishing processes to respond to those unforeseen stories that make dominance so difficult. Clearly, managing the "message" while controlling the necessary new media "means" represent critical challenges to that effort. Nor should one be lulled into believing that the exploitation of new media as a military and strategic enabler is limited to insurgency operations.

While insurgents use new media to affect perceptions, attitudes and beliefs (the cognitive dimension of the information environment) nation states most recently have acted to exploit dependence on new media (specifically the internet) in the physical dimension, revealing an important vulnerability. Russia was alleged to have taken down Estonian government, bank and newspaper websites in what may have been a cyber-attack test in May 2007 prior to their combined cyber and kinetic attack on Georgia in 2008.

China's People's Liberation Army (PLA), the most obvious near peer military competitor to the United States, has gained notoriety recently by being implicated in taking down Pentagon internet capability in June 2007. The Chinese are transforming from a mechanized force to an "informationized" force and have stated intentions to use information warfare "as a tool of war (or) as a way to achieve victory without war...." New media is now playing, and will continue to play an increasingly important role in how the United States conducts warfare. How the U.S. military effectively manages an environment dominated by new media, therefore, becomes increasingly important to success in warfare.

Managing the New Media Environment

The U.S. military has dealt with the new media phenomenon in fits and starts. Certainly one could argue that efforts to deal in this environment have largely been reactive rather than proactive. Soldiers using new media in the form of emails, blogs and digital photography revealed vulnerabilities of U.S. systems as well as operational tactics, techniques and procedures from current operations in Iraq and Afghanistan. This led to a top-down reemphasis on operations security, a program that has always existed, but never in the instantaneous world of information exchange, pointing again to both the power and danger of new media capabilities.

The YouTube and blog response efforts outlined previously are encouraging, albeit reactive efforts to manage this environment. One area of new media the military proactively exploits is knowledge management in order to share lessons learned from recent operational experiences. Interestingly, this process was bottom-up driven in the form of a web site independently developed by junior officers known as CompanyCommand.com. While the power in this forum admittedly lies in the dynamics of people and conversations, its potential would not be realized without the global and instantaneous reach provided by the internet.

Reactive responses, however, will never allow the military to manage the information environment of today, nor the increasingly complex information environment of the future. Instead, a proactive approach to fighting in that environment is necessary. This requires a cultural shift in the military mindset but not a doctrinal change in the way military operations are planned. The military has muddied the waters somewhat in that regard. Recognizing the need to compete in the information battle space they have grappled with the concept of "information operations" in doctrinal publications and applied the concept in current operations with mixed results.

Most recently "strategic communication" has entered the military lexicon. Unfortunately these concepts are widely misunderstood by war fighters and often applied as afterthoughts by senior commanders in attempts to fix problems. Consequently, they become inherently reactive and often ineffective. Senior military leaders have grown up in a culture that emphasizes kinetic war fighting skills, both in planning and execution. In order to ease the cultural transition from this world of bombs and bullets to one where information, driven by new media capabilities, is a significant weapon, it is best to work within both the language and planning methodologies inherent in that military cultural upbringing.

Consequently, war fighters should consider and emphasize the "information effects" they wish to achieve in an operation. This focuses efforts on achieving a military objective, something commanders fully understand, and ensures the full range of capabilities available will be used to that end. The clear statement that will drive planners and subordinate units in that regard is the commander's intent. The Commander's Intent, as part of the formal military planning process "articulate(s) the purpose of the campaign being conducted and the...commander's vision of the military end state when military operations are concluded."

It serves as the impetus for operational planning. The key to proactively managing the information environment, then, lies in a clearly stated *information* end state, that is, a description of what the information environment will look like at the end of the military operation. The information end state should consider both the cognitive and physical dimensions of the information environment. The cognitive end state includes the desired perceptions and attitudes of the target audience (e.g. the indigenous population or international community). The physical information endstate includes a description of the new media capabilities of the adversary at the conclusion of the operation.

A properly articulated information endstate will drive both planning and execution of the military operation with sensitivity toward the new media environment. Military courses of action will be analyzed against this vision and subordinate military units will carry out the operation in order to meet the endstate described within the intent. Sensitized to the commander' intent, planners "wargame" the courses of action with that endstate in mind. Consequently, the planners will consider an enemy's expected reaction to a friendly action in terms of the required information endstate.

This will include recognition that a friendly kinetic action could result in an enemy asymmetric information reaction. Planners can then prepare for a counteraction to blunt the enemy information attack or choose an alternate course of action. Additionally, the information endstate will drive *how* subordinate units carry out their mission. Actions send loud and clear messages to a target audience. Where previously a kinetic solution may have been the preferred choice (driven by inherent organizational culture) the information endstate may instead drive the unit toward a different approach that achieves the stated cognitive effect on perceptions, attitudes and behaviors.

The commander is now unburdened by unfamiliar concepts like information operations and strategic communication. To be sure, his planners and staff experts will apply those concepts to achieve the required information effect, but the key is that they will proactively do so. Nor will they limit themselves to information operations and strategic communication capabilities, but will use every available military capability available, in integrated fashion, to achieve the effect in support of accomplishing the military objective.

The information environment of today guarantees that "wildcards" will present themselves as unpredictable, disruptive forces to current operations. These incidents can significantly impact a military operation, whether the wildcard is the release of Abu Ghraib imagery...or terrorist internet video of gruesome beheadings. While the military response to such circumstances seems necessarily responsive in nature, current planning processes allow proactive consideration of such events as well. In military planning a "branch" is "a contingency option built into the basic plan.... It is used to aid success of the operation based on anticipated events, opportunities, or disruptions caused by enemy actions and reactions.

It answers the question 'what if...?'" Like the commander's intent, however, it requires an organizational culture shift in focus to apply the existing process to the expected new media information environment...but the widely understood process does exist. While branch planning cannot account for every possible wildcard (thus the name) it should anticipate that wildcards will occur and, at a minimum, establish procedures to deal with them.

The Battle of Jenin serves as an example of what could have been if IDF organizational culture had prescribed that the information environment be addressed within the planning process. If the IDF Commander's Intent had articulated an endstate that addressed the cognitive dimension of the information environment would the battle have still resulted in strategic failure? Consider the effect of a stated military endstate where the people of Jenin would "remain neutral in their attitudes toward Israel" and "the international community would understand and support IDF efforts to defeat terrorism." Given that simple statement that drives planning and execution it is likely that the media would have been allowed to embed with the IDF during the battle in order to allow Israel to compete for international legitimacy.

It's also likely that subordinate units would have carried out operations by using tactics other than bulldozing buildings to achieve their objectives in order to maintain Palestinian neutrality. Finally, branches to consider wildcards could have been developed that defined a process to address the now expected mis- and dis-information that the numerous cell phone interviews coming out of the refugee camps engendered.

Those counter-wildcard procedures could have required IDF forces to carry helmet cameras to film operations to prove proportional response by the IDF and to immediately counter any framed or altered imagery distributed by the insurgents. Given the importance of new media and the information environment, a statement of the information endstate must be required within commander's intent. This simple change in doctrinal requirement, supported by the military education system can drive organizational culture change and ensure the military proactively manages the information environment.

There is a generation gap between senior war fighters and their junior officers…a gap defined by digital immigrants and digital natives. The junior officers who developed CompanyCommand.com, as digital natives, fully appreciate the importance and power of new media and proactively exploit its capabilities. Senior war fighters, on the other hand, certainly understand its importance but lack the cultural upbringing to see it in the context of current military operations. Consequently, military operations are often impacted by unanticipated consequences enabled by the use of new media in today's information environment.

Closing this gap to achieve success requires a cultural shift in the minds of these senior leaders. Introducing new concepts, like information operations and strategic communication, while academically interesting and militarily prudent, does not enable the required cultural enlightenment. Instead, as "new" concepts they lie outside the generational culture of these seniors who then often relegate the information fight to specialized staffs established to understand those espoused doctrines and capabilities.

Recent proclamations by senior military leaders are encouraging regarding closing this generational gap. However, if the United States military hopes to fight and win in a future information environment dominated by new media then that gap must be closed as quickly as possible. Leveraging new media to achieve military objectives and defeat an adversary's skilled use of it requires a significant cultural shift at these senior ranks, but also recognizes that current planning processes remain valid. As Torie Clarke, former Pentagon press spokesperson, noted regarding information and its impact: "It comes down to people, planning and integration." Implementing change within that construct is the easiest way to effectively impact a culture where managing the information environment is not only achievable, but must be expected.

3

Media Coverage and Terrorist Objectives

Kellner synopsizes a key assertion of the media-terrorism connection, that Osama bin Laden and al Qaida use the spectacle of terror to promote its agenda in a media-saturated era. Kellner also addresses the counter-terrorism response, positing that both Bush Administrations deployed the terror spectacle to promote their geo-political ends in the Middle East. Al Qaida's September 11 attacks continued terrorist groups' practice of staging extravagant terror spectacles to gain worldwide attention and dramatize their causes, in part through the diffusion effect of fear and uncertainty among targeted populations. The 9/11 attacks were unique, however, for their novelty of target and means, such that the drama of the attack and the response remain unparalleled.

Nacos likewise speaks to the "spectacle" of terrorist attacks: "In a popular culture inundated with images of violence, Americans could not comprehend what was happening before their eyes and what had happened already. The horror of the quadruple hijack and suicide coup was as real as in a movie, but was surreal in life." The coverage of the attacks served to displace the "terrorism as theater" metaphor with that of television spectacular, and reinforced al Qaida's desire for a perfectly choreographed production aimed at American and international audiences.

While acknowledging that acts of terrorism do not require extensive media coverage to satisfy all of the goals and objectives of the perpetrators, Traugott and Brader illustrate how the media and media coverage can be considered modern tools of the terrorists, by magnifying the size of the audience and exposure to the consequences. Like Kellner, Traugott and Brader argue that diffusion effects of coverage increase social and political concerns and anxiety, which are often directed toward governmental officials and institutions. In explaining news practices and the general consistency in news organizations' approach, Traugott and Brader's study takes into consideration the media models of selection and prominence: the standard criteria of newsworthiness applies, namely timeliness, proximity, impact, conflict, sensationalism and novelty.

Wittebols adds that media coverage of terrorism is informed by the role of social power. Wittebols cites in particular a terrorist incident's relationship to powerful elements in society, the relationship of the media to same, and the media's institutional role of shaping audience world views. Accordingly, the net effects of media coverage are contingent on whether terrorist acts challenge or reinforce the institutional bases of power to which the media are related and rely upon for sourcing.

In the essay, Wittebols introduces a new construct useful for media analyses that take the rhetoric-and-power point of view, which is the contrast between *grievance terrorism* and *institutional terrorism*. In the former, government sources that dominate news set an agenda hostile to groups that challenge the United States or Western interests. In the latter, a lack of coverage conceals the degree of U.S. complicity through support of repressive governments. Wittebols argues that the amount of coverage given to both grievance and institutional terrorism reveals much about the media's relationship to domestic governments, a primary factor being the role of sources used to set the tone and flavor of coverage.

This institutional bias contributes to three themes emergent in media coverage: we are victims – they are terrorists; the United States strives to "do good" in the world; and terrorism is the product of irrational minds rather than objective conditions. Labeling is among the linguistic cues used to recognize the political, cultural and ideological strains emergent in news coverage. Steuter focused in-depth on the use of labels in an analysis of *Time* magazine coverage of terrorism, published between January and December 1986. Steuter documented and delineated the way in which ideology operates in the presentation of terrorism news.

His objectives were to chart the function of "terrorist" and "freedom fighter" labels vis-à-vis group legitimacy; to show that the social and professional assumptions producing frames of reference are not neutral; and to examine the resulting long-term consequences of language use. Steuter's analysis confirmed that labeling confers varying degrees of legitimacy. Steuter compared five different patterns of labeling, categorized as: "terrorist/ terrorism" (at the time a narrow view of actions deserving the term); "unpopular insurgency movement" (focusing on violence and destruction); "popular insurgency movement" (sympathetic treatment, status as "freedom fighters"); factional violence and political in-fighting (critical of the disorder caused by rival groups); and state terrorism (redefining official violence of U.S. allies to avoid ascription of the foregoing labels).

In employing these typologies, the media invoked a second ideological mechanism, that of the underworld, to describe terrorists as "shadowy" or cloaked in mystery. While *Time* magazine coverage conveyed the "official perspective" on terrorism (almost every article examined cited one or more authoritative sources), the coverage also employed balance of coverage to sustain the appearance of neutrality. Furthermore, the coverage was both episodic, removing specific terrorist acts from political and social explanation, and generalized, comparing the phenomenon of terrorism to natural forces or disasters; the terrorist actors were portrayed as irrational and barbarous. As a result, coverage tended to amplify violence and exaggerate indiscriminant targeting, leaving readers with a greater sense of fear and concern than objectively warranted.

Approaches to Terrorism Analysis

The Multicausal Approach

Terrorism usually results from multiple causal factors— not only psychological but also economic, political, religious, and sociological factors, among others. There is even an hypothesis that it is caused by physiological factors, as discussed below. Because terrorism is a multicausal phenomenon, it would be simplistic and erroneous to explain an act of terrorism by a single cause, such as the psychological need of the terrorist to perpetrate an act of violence.

For Paul Wilkinson (1977), the causes of revolution and political violence in general are also the causes of terrorism. These include ethnic conflicts, religious and ideological conflicts, poverty, modernization stresses, political inequities, lack of peaceful communications channels, traditions of violence, the existence of a revolutionary group, governmental weakness and ineptness, erosions of confidence in a regime, and deep divisions within governing elites and leadership groups.

The Political Approach

The alternative to the hypothesis that a terrorist is born with certain personality traits that destine him or her to become a terrorist is that the root causes of terrorism can be found in influences emanating from environmental factors. Environments conducive to the rise of terrorism include international and national environments, as well as sub national ones such as universities, where many terrorists first become familiar with Marxist-Leninist ideology or other revolutionary ideas and get involved with radical groups. Russell and Miller identify universities as the major recruiting ground for terrorists.

Having identified one or more of these or other environments, analysts may distinguish between precipitants that started the outbreak of violence, on the one hand, and preconditions that allowed the precipitants to instigate the action, on the other hand. Political scientists Chalmers Johnson (1978) and Martha Crenshaw (1981) have further subdivided preconditions into permissive factors, which engender a terrorist strategy and make it attractive to political dissidents, and direct situational factors, which motivate terrorists.

Permissive causes include urbanization, the transportation system (for example, by allowing a terrorist to quickly escape to another country by taking a flight), communications media, weapons availability, and the absence of security measures. An example of a situational factor for Palestinians would be the loss of their homeland of Palestine. Various examples of international and national or sub national theories of terrorism can be cited.

An example of an international environment hypothesis is the view proposed by Brian M. Jenkins (1979) that the failure of rural guerrilla movements in Latin America pushed the rebels into the cities. (This hypothesis, however, overlooks the national causes of Latin American terrorism and fails to explain why rural guerrilla movements continue to thrive in Colombia.) Jenkins also notes that the defeat of Arab armies in the 1967 Six-Day War caused the Palestinians to abandon hope for a conventional military solution to their problem and to turn to terrorist attacks.

The Organizational Approach

Some analysts, such as Crenshaw (1990: 250), take an organization approach to terrorism and see terrorism as a rational strategic course of action decided on by a group. In her view, terrorism is not committed by an individual. Rather, she contends that "Acts of terrorism are committed by groups who reach collective decisions based on commonly held beliefs, although the level of individual commitment to the group and its beliefs varies."

Crenshaw has not actually substantiated her contention with case studies that show how decisions are supposedly reached collectively in terrorist groups. That kind of inside information, to be sure, would be quite difficult to obtain without a former decision-maker within a terrorist group providing it in the form of a published autobiography or an interview, or even as a paid police informer. Crenshaw may be partly right, but her organizational approach would seem to be more relevant to guerrilla organizations that are organized along traditional Marxist-Leninist lines, with a general secretariat headed by a secretary general, than to terrorist groups per se.

The FARC, for example, is a guerrilla organization, albeit one that is not averse to using terrorism as a tactic. The six members of the FARC's General Secretariat participate in its decision-making under the overall leadership of Secretary General Manuel Marulanda Vélez. The hardline military leaders, however, often exert disproportionate influence over decision-making. Bona fide terrorist groups, like cults, are often totally dominated by a single individual leader, be it Abu Nidal, Ahmed Jibril, Osama bin Laden, or Shoko Asahara.

It seems quite improbable that the terrorist groups of such dominating leaders make their decisions collectively. By most accounts, the established terrorist leaders give instructions to their lieutenants to hijack a jetliner, assassinate a particular person, bomb a U.S. Embassy, and so forth, while leaving operational details to their lieutenants to work out. The top leader may listen to his lieutenants' advice, but the top leader makes the final decision and gives the orders.

The Physiological Approach

The physiological approach to terrorism suggests that the role of the media in promoting the spread of terrorism cannot be ignored in any discussion of the causes of terrorism. Thanks to media coverage, the methods, demands, and goals of terrorists are quickly made known to potential terrorists, who may be inspired to imitate them upon becoming stimulated by media accounts of terrorist acts.

The diffusion of terrorism from one place to another received scholarly attention in the early 1980s. David G. Hubbard (1983) takes a physiological approach to analyzing the causes of terrorism. He discusses three substances produced in the body under stress: norepinephrine, a compound produced by the adrenal gland and sympathetic nerve endings and associated with the "fight or flight" physiological response of individuals in stressful situations; acetylcholine, which is produced by the parasympathetic nerve endings and acts to dampen the accelerated norepinephrine response; and endorphins, which develop in the brain as a response to stress and "narcotize" the brain, being 100 times more powerful than morphine. Because these substances occur in the terrorist, Hubbard concludes that much terrorist violence is rooted not in the psychology but in the physiology of the terrorist, partly the result of "stereotyped, agitated tissue response" to stress. Hubbard's conclusion suggests a possible explanation for the spread of terrorism, the so-called contagion effect.

Kent Layne Oots and Thomas C. Wiegele (1985) have

also proposed a model of terrorist contagion based on physiology. Their model demonstrates that the psychological state of the potential terrorist has important implications for the stability of society. In their analysis, because potential terrorists become aroused in a violence-accepting way by media presentations of terrorism, "Terrorists must, by the nature of their actions, have an attitude which allows violence." One of these attitudes, they suspect, may be Machiavellianism because terrorists are disposed to manipulating their victims as well as the press, the public, and the authorities. They note that the potential terrorist "need only see that terrorism has worked for others in order to become aggressively aroused."

According to Oots and Wiegele, an individual moves from being a potential terrorist to being an actual terrorist through a process that is psychological, physiological, and political. "If the neurophysiological model of aggression is realistic," Oots and Wiegele assert, "there is no basis for the argument that terrorism could be eliminated if its sociopolitical causes were eliminated." They characterize the potential terrorist as "a frustrated individual who has become aroused and has repeatedly experienced the fight or flight syndrome. Moreover, after these repeated arousals, the potential terrorist seeks relief through an aggressive act and also seeks, in part, to remove the initial cause of his frustration by achieving the political goal which he has hitherto been denied."

D. Guttman (1979) also sees terrorist actions as being aimed more at the audience than at the immediate victims. It is, after all, the audience that may have to meet the terrorist's demands. Moreover, in Guttman's analysis, the terrorist requires a liberal rather than a right-wing audience for success. Liberals make the terrorist respectable by accepting the ideology that the terrorist alleges informs his or her acts. The terrorist also requires liberal control of the media for the transmission of his or her ideology.

The Psychological Approach

In contrast with political scientists and sociologists, who are interested in the political and social contexts of terrorist groups, the relatively few psychologists who study terrorism are primarily interested in the micro-level of the individual terrorist or terrorist group. The psychological approach is concerned with the study of terrorists per se, their recruitment and induction into terrorist groups, their personalities, beliefs, attitudes, motivations, and careers as terrorists.

Hypotheses of Terrorism

If one accepts the proposition that political terrorists are made, not born, then the question is what makes a terrorist. Although the scholarly literature on the psychology of terrorism is lacking in full-scale, quantitative studies from which to ascertain trends and develop general theories of terrorism, it does appear to focus on several theories. One, the Olson hypothesis, suggests that participants in revolutionary violence predicate their behavior on a rational cost-benefit calculus and the conclusion that violence is the best available course of action given the social conditions. The notion that a group rationally chooses a terrorism strategy is questionable, however. Indeed, a group's decision to resort to terrorism is often divisive, sometimes resulting in factionalization of the group.

Frustration-Aggression Hypothesis

The frustration-aggression hypothesis of violence is prominent in the literature. This hypothesis is based mostly on the relative-deprivation hypothesis, as proposed by Ted Robert Gurr (1970), an expert on violent behaviors and movements, and reformulated by J.C. Davies (1973) to include a gap between rising expectations and need satisfaction. Another proponent of this hypothesis, Joseph Margolin, argues that "much terrorist behavior is a response to the frustration of various political, economic, and personal needs or objectives."

Other scholars, however have dismissed the frustration-aggression hypothesis as simplistic, based as it is on the erroneous assumption that aggression is always a consequence of frustration. According to Franco Ferracuti (1982), a University of Rome professor, a better approach than these and other hypotheses, including the Marxist theory, would be a sub cultural theory, which takes into account that terrorists live in their own subculture, with their own value systems.

Similarly, political scientist Paul Wilkinson (1974: 127) faults the frustration-aggression hypothesis for having "very little to say about the social psychology of prejudice and hatred..." and fanaticisms that "play a major role in encouraging extreme violence." He believes that "Political terrorism cannot be understood outside the context of the development of terroristic, or potentially terroristic, ideologies, beliefs and life-styles."

Negative Identity Hypothesis

Using Erikson's theory of identity formation, particularly his concept of negative identity, the late political psychologist Jeanne N. Knutson (1981) suggests that the political terrorist consciously assumes a negative identity. One of her examples is a Croatian terrorist who, as a member of an oppressed ethnic minority, was disappointed by the failure of his aspiration to attain a university education, and as a result assumed a negative identity by becoming a terrorist.

Negative identity involves a vindictive rejection of the role regarded as desirable and proper by an individual's family and community. In Knutson's view, terrorists engage in terrorism as a result of feelings of rage and helplessness over the lack of alternatives. Her political science-oriented viewpoint seems to coincide with the frustration-aggression hypothesis.

Narcissistic Rage Hypothesis

The advocates of the narcissism-aggression hypothesis include psychologists Jerrold M. Post, John W. Crayton, and Richard M. Pearlstein. Taking the-terrorists-as-mentally-ill approach, this hypothesis concerns the early development of the terrorist. Basically, if primary narcissism in the form of the "grandiose self" is not neutralized by reality testing, the grandiose self produces individuals who are sociopathic, arrogant, and lacking in regard for others. Similarly, if the psychological form of the "idealized parental ego" is not neutralized by reality testing, it can produce a condition of helpless defeatism, and narcissistic defeat can lead to reactions of rage and a wish to destroy the source of narcissistic injury.

"As a specific manifestation of narcissistic rage, terrorism occurs in the context of narcissistic injury," writes Crayton. For Crayton, terrorism is an attempt to acquire or maintain power or control by intimidation. He suggests that the "meaningful high ideals" of the political terrorist group "protect the group members from experiencing shame."

In Post's view, a particularly striking personality trait of people who are drawn to terrorism "is the reliance placed on the psychological mechanisms of "externalization" and 'splitting'." These are psychological mechanisms, he explains, that are found in "individuals with narcissistic and borderline personality disturbances." "Splitting," he explains, is a mechanism characteristic of people whose personality development is shaped by a particular type of psychological damage (narcissistic injury) during childhood. Those individuals with a damaged self-concept have failed to integrate the good and bad parts of the self, which are instead split into the "me" and the "not me."

These individuals, who have included Hitler, need an outside enemy to blame for their own inadequacies and weaknesses. The data examined by Post, including a 1982 West German study, indicate that many terrorists have not been successful in their personal, educational, and vocational lives. Thus, they are drawn to terrorist groups, which have an us-versus-them outlook. This hypothesis, however, appears to be contradicted by the increasing number of terrorists who are well-educated professionals, such as chemists, engineers, and physicists.

The psychology of the self is clearly very important in understanding and dealing with terrorist behavior, as in incidents of hostage-barricade terrorism. Crayton points out that humiliating the terrorists in such situations by withholding food, for example, would be counter-productive because "the very basis for their activity stems from their sense of low self-esteem and humiliation."

Using a Freudian analysis of the self and the narcissistic personality, Pearlstein (1991) eruditely applies the psychological concept of narcissism to terrorists. He observes that the political terrorist circumvents the psycho political liabilities of accepting himself or herself as a terrorist with a negative identity through a process of rhetorical self-justification that is reinforced by the group's group-think. His hypothesis, however, seems too speculative a construct to be used to analyze terrorist motivation independently of numerous other factors.

For example, politically motivated hijackers have rarely acted for self-centered reasons, but rather in the name of the political goals of their groups. It also seems questionable that terrorist suicide-bombers, who deliberately sacrificed themselves in the act, had a narcissistic personality.

Terrorist Motivation

In addition to drawing on political science and sociology, this study draws on the discipline of psychology, in an attempt to explain terrorist motivation and to answer questions such as who become terrorists and what kind of individuals join terrorist groups and commit public acts of shocking violence. Although there have been numerous attempts to explain terrorism from a psychiatric or psychological perspective, Wilkinson notes that the psychology and beliefs of terrorists have been inadequately explored.

Most psychological analyses of terrorists and terrorism, according to psychologist Maxwell Taylor (1988), have attempted to address what motivates terrorists or to describe personal characteristics of terrorists, on the assumption that terrorists can be identified by these attributes. However, although an understanding of the terrorist mindset would be the key to understanding how and why an individual becomes a terrorist, numerous psychologists have been unable to adequately define it. Indeed, there appears to be a general agreement among psychologists who have studied the subject that there is no one terrorist mindset. This view, however, itself needs to be clarified.

The topic of the terrorist mindset was discussed at a Rand conference on terrorism coordinated by Brian M. Jenkins in September 1980. The observations made about terrorist mindsets at that conference considered individuals, groups, and individuals as part of a group. The discussion revealed how little was known about the nature of terrorist mindsets, their causes and consequences, and their significance for recruitment, ideology, leader-follower relations, organization, decision making about targets and tactics, escalation of violence, and attempts made by disillusioned terrorists to exit from the terrorist group.

Although the current study has examined these aspects of the terrorist mindset, it has done so within the framework of a more general tasking requirement. Additional research and analysis would be needed to focus more closely on the concept of the terrorist mindset and to develop it into a more useful method for profiling terrorist groups and leaders on a more systematic and accurate basis. Within this field of psychology, the personality dynamics of individual terrorists, including the causes and motivations behind the decision to join a terrorist group and to commit violent acts, have also received attention.

Other small-group dynamics that have been of particular interest to researchers include the terrorists' decision-making patterns, problems of leadership and authority, target selection, and group mindset as a pressure tool on the individual. Attempts to explain terrorism in purely psychological terms ignore the very real economic, political, and social factors that have always motivated radical activists, as well as the possibility that biological or physiological variables may play a role in bringing an individual to the point of perpetrating terrorism.

Although this study provides some interdisciplinary context to the study of terrorists and terrorism, it is concerned primarily with the socio-psychological approach. Knutson (1984), Executive Director of the International Society of Political Psychology until her death in 1982, carried out an extensive international research project on the psychology of political terrorism. The basic premise of terrorists whom she evaluated in depth was "that their violent acts stem from feelings of rage and hopelessness engendered by the belief that society permits no other access to information-dissemination and policy-formation processes."

The social psychology of political terrorism has received extensive analysis in studies of terrorism, but the individual psychology of political and religious terrorism has been largely ignored. Relatively little is known about the terrorist as an individual, and the psychology of terrorists remains poorly understood, despite the fact that there have been a number of individual biographical accounts, as well as sweeping sociopolitical or psychiatric generalizations.

A lack of data and an apparent ambivalence among many academic researchers about the academic value of terrorism research have contributed to the relatively little systematic social and psychological research on terrorism. This is unfortunate because psychology, concerned as it is with behavior and the factors that influence and control behavior, can provide practical as opposed to conceptual knowledge of terrorists and terrorism. A principal reason for the lack of psychometric studies of terrorism is that researchers have little, if any, direct access to terrorists, even imprisoned ones.

Occasionally, a researcher has gained special access to a terrorist group, but usually at the cost of compromising the credibility of her/her research. Even if a researcher obtains permission to interview an incarcerated terrorist, such an interview would be of limited value and reliability for the purpose of making generalizations. Most terrorists, including imprisoned ones, would be loath to reveal their group's operational secrets to their interrogators, let alone to journalists or academic researchers, whom the terrorists are likely to view as representatives of the "system" or perhaps even as intelligence agents in disguise.

Even if terrorists agree to be interviewed in such circumstances, they may be less than candid in answering questions. For example, most imprisoned Red Army Faction members reportedly declined to be interviewed by West German social scientists. Few researchers or former terrorists write exposés of terrorist groups. Those who do could face retaliation. For example, the LTTE shot to death an anti-LTTE activist, Sabaratnam Sabalingam, in Paris on May 1, 1994, to prevent him from publishing an anti-LTTE book. The LTTE also murdered Dr. Rajani Thiranagama, a Tamil, and one of the four Sri Lankan authors of *The Broken Palmyrah*, which sought to examine the "martyr" cult.

Individuals who become terrorists often are unemployed, socially alienated individuals who have dropped out of society. Those with little education, such as youths in Algerian ghettos or the Gaza Strip, may try to join a terrorist group out of boredom and a desire to have an action-packed adventure in pursuit of a cause they regard as just. Some individuals may be motivated mainly by a desire to use their special skills, such as bomb-making. The more educated youths may be motivated more by genuine political or religious convictions. The person who becomes a terrorist in Western countries is generally both intellectual and idealistic. Usually, these disenchanted youths, both educated or uneducated, engage in occasional protest and dissidence.

Potential terrorist group members often start out as sympathizers of the group. Recruits often come from support organizations, such as prisoner support groups or student activist groups. From sympathizer, one moves to passive supporter. Often, violent encounters with police or other security forces motivate an already socially alienated individual to join a terrorist group. Although the circumstances vary, the end result of this gradual process is that the individual, often with the help of a family member or friend with terrorist contacts, turns to terrorism. Membership in a terrorist group, however, is highly selective.

Over a period as long as a year or more, a recruit generally moves in a slow, gradual fashion toward full membership in a terrorist group. An individual who drops out of society can just as well become a monk or a hermit instead of a terrorist. For an individual to choose to become a terrorist, he or she would have to be motivated to do so. Having the proper motivation, however, is still not enough. The would-be terrorist would need to have the opportunity to join a terrorist group.

And like most job seekers, he or she would have to be acceptable to the terrorist group, which is a highly exclusive group. Thus, recruits would not only need to have a personality that would allow them to fit into the group, but ideally a certain skill needed by the group, such as weapons or communications skills. The psychology of joining a terrorist group differs depending on the typology of the group. Someone joining an anarchistic or a Marxist-

Leninist terrorist group would not likely be able to count on any social support, only social opprobrium, whereas someone joining an ethnic separatist group like ETA or the IRA would enjoy considerable social support and even respect within ethnic enclaves.

Psychologist Eric D. Shaw (1986:365) provides a strong case for what he calls "The Personal Pathway Model", by which terrorists enter their new profession. The components of this pathway include early socialization processes; narcissistic injuries; escalatory events, particularly confrontation with police; and personal connections to terrorist group members, as follows:

The personal pathway model suggests that terrorists came from a selected, at risk population, who have suffered from early damage to their self-esteem. Their subsequent political activities may be consistent with the liberal social philosophies of their families, but go beyond their perception of the contradiction in their family's beliefs and lack of social action. Family political philosophies may also serve to sensitize these persons to the economic and political tensions inherent throughout modern society. As a group, they appear to have been unsuccessful in obtaining a desired traditional place in society, which has contributed to their frustration. The underlying need to belong to a terrorist group is symptomatic of an incomplete or fragmented psychosocial identity. (In Kohut's terms—a defective or fragmented "group self"). Interestingly, the acts of security forces or police are cited as provoking more violent political activity by these individuals and it is often a personal connection to other terrorists which leads to membership in a violent group (shared external targets).

Increasingly, terrorist organizations in the developing world are recruiting younger members. The only role models for these young people to identify with are often terrorists and guerrillas. Abu Nidal, for example, was able to recruit alienated, poor, and uneducated youths thrilled to be able to identify themselves with a group led by a well-known but mysterious figure.

During the 1980s and early 1990s, thousands of foreign Muslim volunteers (14,000, according to *Jane's Intelligence Review*) — angry, young, and zealous and from many countries, including the United States — flocked to training camps in Afghanistan or the Pakistan-Afghan border region to learn the art of combat. They ranged in age from 17 to 35. Some had university educations, but most were uneducated, unemployed youths without any prospects.

Deborah M. Galvin (1983) notes that a common route of entry into terrorism for female terrorists is through political involvement and belief in a political cause. The Intifada (see Glossary), for example, radicalized many young Palestinians, who later joined terrorist organizations. At least half of the Intifada protesters were young girls. Some women are recruited into terrorist organizations by boyfriends. A significant feature that Galvin feels may characterize the involvement of the female terrorist is the "male or female lover/female accomplice ... scenario."

The lover, a member of the terrorist group, recruits the female into the group. One ETA female member, "Begona," told Eileen MacDonald (1992) that was how she joined at age 25: "I got involved [in ETA] because a man I knew was a member."

A woman who is recruited into a terrorist organization on the basis of her qualifications and motivation is likely to be treated more professionally by her comrades than one who is perceived as lacking in this regard. Two of the PFLP hijackers of Sabena Flight 517 from Brussels to Tel Aviv on May 8, 1972, Therese Halsa, 19, and Rima Tannous, 21, had completely different characters.

Therese, the daughter of a middle-class Arab family, was a nursing student when she was recruited into Fatah by a fellow student and was well regarded in the organization. Rima, an orphan of average intelligence, became the mistress of a doctor who introduced her to drugs and recruited her into Fatah. She became totally dependent on some Fatah members, who subjected her to physical and psychological abuse.

Various terrorist groups recruit both female and male members from organizations that are lawful. For example, ETA personnel may be members of Egizan ("Act Woman"), a feminist movement affiliated with ETA's political wing; the Henri Batasuna (Popular Unity) party; or an amnesty group seeking release for ETA members. While working with the amnesty group, a number of women reportedly tended to become frustrated over mistreatment of prisoners and concluded that the only solution was to strike back, which they did by joining the ETA. "Women seemed to become far more emotionally involved than men with the suffering of prisoners," an ETA member, "Txikia," who joined at age 20, told MacDonald, "and when they made the transition from supporter to guerrilla, appeared to carry their deeper sense of commitment with them into battle."

A common stereotype is that someone who commits such abhorrent acts as planting a bomb on an airliner, detonating a vehicle bomb on a city street, or tossing a grenade into a crowded sidewalk café is abnormal. The psychopathological orientation has dominated the psychological approach to the terrorist's personality. As noted by Taylor, two basic psychological approaches to understanding terrorists have been commonly used: the terrorist is viewed either as mentally ill or as a fanatic. For Walter Laqueur, "Terrorists are fanatics and fanaticism frequently makes for cruelty and sadism."

This study is not concerned with the lone terrorist, such as the Unabomber in the United States, who did not belong to any terrorist group. Criminologist Franco Ferracuti has noted that there is "no such thing as an isolated terrorist— that's a mental case." Mentally unbalanced individuals have been especially attracted to airplane hijacking. David G. Hubbard (1971) conducted a psychiatric study of airplane hijackers in 1971 and concluded that skyjacking is used by psychiatrically ill patients as an expression of illness.

His study revealed that skyjackers shared several common traits: a violent father, often an alcoholic; a deeply religious mother, often a religious zealot; a sexually shy, timid, and passive personality; younger sisters toward whom the skyjackers acted protectively; and poor achievement, financial failure, and limited earning potential. Those traits, however, are shared by many people who do not hijack airplanes. Thus, profiles of mentally unstable hijackers would seem to be of little, if any, use in detecting a potential hijacker in advance.

A useful profile would probably have to identify physical or behavioral traits that might alert authorities to a potential terrorist before a suspect is allowed to board an aircraft, that is, if hijackers have identifiable personality qualities. In the meantime, weapons detection, passenger identification, and onboard security guards may be the only preventive measures. Even then, an individual wanting to hijack an airplane can often find a way.

Japan's Haneda Airport screening procedures failed to detect a large knife that a 28-year-old man carried aboard an All Nippon Airways jumbo jet on July 23, 1999, and used to stab the pilot (who died) and take the plane's controls until overpowered by others. Although police have suggested that the man may have psychiatric problems, the fact that he attempted to divert the plane to the U.S. Yokota Air Base north of Tokyo, at a time when the airbase was a subject of controversy because the newly elected governor of Tokyo had demanded its closure, suggests that he may have had a political or religious motive.

There have been cases of certifiably mentally ill terrorists. Klaus Jünschke, a mental patient, was one of the most ardent members of the Socialist Patients' Collective (SPK), a German terrorist group working with the Baader-Meinhof Gang. In some instances, political terrorists have clearly exhibited psychopathy. For example, in April 1986 Nezar Hindawi, a freelance Syrian-funded Jordanian terrorist and would-be agent of Syrian intelligence, sent his pregnant Irish girlfriend on an El Al flight to Israel, promising to meet her there to be married.

Unknown to her, however, Hindawi had hidden a bomb (provided by the Abu Nidal Organization (ANO)) in a false bottom to her hand luggage. His attempt to bomb the airliner in midair by duping his pregnant girlfriend was thwarted when the bomb was discovered by Heathrow security personnel. Taylor regards Hindawi's behavior in this incident as psychopathic because of Hindawi's willingness to sacrifice his fiancé and unborn child.

Jerrold Post (1990), a leading advocate of the terrorists-as-mentally ill approach, has his own psychological hypothesis of terrorism. Although he does not take issue with the proposition that terrorists reason logically, Post argues that terrorists' reasoning process is characterized by what he terms "terrorist psycho-logic." In his analysis, terrorists do not willingly resort to terrorism as an intentional choice. Rather, he argues that "political terrorists are driven to commit acts of violence as a consequence of psychological forces, and that their special psycho-logic is constructed to rationalize acts they are psychologically compelled to commit"(1990:25).

Post's hypothesis that terrorists are motivated by psychological forces is not convincing and seems to ignore the numerous factors that motivate terrorists, including their ideological convictions. Post (1997) believes that the most potent form of terrorism stems from those individuals who are bred to hate, from generation to generation, as in Northern Ireland and the Basque country.

For these terrorists, in his view, rehabilitation in nearly impossible because ethnic animosity or hatred is "in their blood" and passed from father to son. Post also draws an interesting distinction between "anarchic-ideologues" such as the Italian Red Brigades (Brigate Rosse) and the German RAF (aka the Baader-Meinhof Gang), and the "nationalist-separatist" groups such as the ETA, or the IRA, stating that:

There would seem to be a profound difference between terrorists bent on destroying their own society, the "world of their fathers," and those whose terrorist activities carry on the mission of their fathers. To put it in other words, for some, becoming terrorists is an act of retaliation for real and imagined hurts *against the society of their parents*; for others, it is an act of retaliation against society *for the hurt done to their parents.*... This would suggest more conflict, more psychopathology, among those committed to anarchy and destruction of society.... (1984:243)

Indeed, author Julian Becker (1984) describes the German terrorists of the Baader-Meinhof Gang as "children without fathers." They were sons and daughters of fathers who had either been killed by Nazis or survived Nazism. Their children despised and rebelled against them because of the shame of Nazism and a defeated Germany. One former RAF female member told MacDonald: "We hated our parents because they were former Nazis, who had never come clean about their past." Similarly, Gunther Wagenlehner (1978:201) concludes that the motives of RAF terrorists were unpolitical and belonged "more to the area of psychopathological disturbances." Wagenlehner found that German terrorists blamed the government for failing to solve their personal problems.

Not only was becoming a terrorist "an individual form of liberation" for radical young people with personal problems, but "These students became terrorists because they suffered from acute fear and from aggression and the masochistic desire to be pursued." In short, according to

Wagenlehner, the West German anarchists stand out as a major exception to the generally non pathological characteristics of most terrorists.

Psychologist Konrad Kellen (1990:43) arrives at a similar conclusion, noting that most of the West German terrorists "suffer from a deep psychological trauma" that "makes them see the world, including their own actions and the expected effects of those actions, in a grossly unrealistic light" and that motivates them to kill people. Sociologist J. Bowyer Bell (1985) also has noted that European anarchists, unlike other terrorists, belong more to the "province of psychologists than political analysts...."

Post's distinction between anarchic-ideologues and ethnic separatists appears to be supported by Rona M. Fields's (1978) psychometric assessment of children in Northern Ireland. Fields found that exposure to terrorism as a child can lead to a proclivity for terrorism as an adult. Thus, a child growing up in violence-plagued West Belfast is more likely to develop into a terrorist as an adult than is a child growing up in peaceful Oslo, Norway, for example.

Maxwell Taylor, noting correctly that there are numerous other factors in the development of a terrorist, faults Fields's conclusions for, among other things, a lack of validation with adults. Maxwell Taylor overlooks, however, that Field's study was conducted over an eight-year period. Taylor's point is that Field's conclusions do not take into account that relatively very few children exposed to violence, even in Northern Ireland, grow up to become terrorists.

A number of other psychologists would take issue with another of Post's contentions—that the West German anarchists were more pathological than Irish terrorists. For example, psychiatrist W. Rasch (1979), who interviewed a number of West German terrorists, determined that "no conclusive evidence has been found for the assumption that a significant number of them are disturbed or abnormal."

For Rasch the argument that terrorism is pathological behavior only serves to minimize the political or social issues that motivated the terrorists into action. And psychologist Ken Heskin (1984), who has studied the psychology of terrorism in Northern Ireland, notes that "In fact, there is no psychological evidence that terrorists are diagnosably psychopathic or otherwise clinically disturbed."

Although there may have been instances in which a mentally ill individual led a terrorist group, this has generally not been the case in international terrorism. Some specialists point out, in fact, that there is little reliable evidence to support the notion that terrorists in general are psychologically disturbed individuals. The careful, detailed planning and well-timed execution that have characterized many terrorist operations are hardly typical of mentally disturbed individuals.

There is considerable evidence, on the contrary, that international terrorists are generally quite sane. Crenshaw (1981) has concluded from her studies that "the outstanding common characteristic of terrorists is their normality." This view is shared by a number of psychologists. For example, C.R. McCauley and M.E. Segal (1987) conclude in a review of the social psychology of terrorist groups that "the best documented generalization is negative; terrorists do not show any striking psychopathology." Heskin (1984) did not find members of the IRA to be emotionally disturbed. It seems clear that terrorists are extremely alienated from society, but alienation does not necessarily mean being mentally ill.

Maxwell Taylor (1984) found that the notion of mental illness has little utility with respect to most terrorist actions. Placing the terrorist within the ranks of the mentally ill, he points out, makes assumptions about terrorist motivations and places terrorist behavior outside the realms of both the normal rules of behavior and the normal process of law. He points out several differences that separate the psychopath from the political terrorist, although the two may not be mutually exclusive, as in the case of Hindawi. One difference is the psychopath's inability to profit from experience.

Another important difference is that, in contrast to the terrorist, the purposefulness, if any, of a psychopath's actions is personal. In addition, psychopaths are too unreliable and incapable of being controlled to be of use to terrorist groups. Taylor notes that terrorist groups need discreet activists who do not draw attention to themselves and who can merge back into the crowd after executing an operation. For these reasons, he believes that "it may be inappropriate to think of the terrorist as mentally ill in conventional terms" (1994:92).

Taylor and Ethel Quayle (1994:197) conclude that "the active terrorist is not discernibly different in psychological terms from the non-terrorist." In other words, terrorists are recruited from a population that describes most of us. Taylor and Quayle also assert that "in psychological terms, there are no special qualities that characterize the terrorist." Just as there is no necessary reason why people sharing the same career in normal life necessarily have psychological characteristics in common, the fact that terrorists have the same career does not necessarily mean that they have anything in common psychologically.

The selectivity with which terrorist groups recruit new members helps to explain why so few pathologically ill individuals are found within their ranks. Candidates who appear to be potentially dangerous to the terrorist group's survival are screened out. Candidates with unpredictable or uncontrolled behavior lack the personal attributes that the terrorist recruiter is looking for.

Many observers have noted that the personality of the terrorist has a depressive aspect to it, as reflected in the terrorist's death-seeking or death-confronting behavior. The terrorist has often been described by psychologists as incapable of enjoying anything (an hedonic) or forming meaningful interpersonal relationships on a reciprocal level.

According to psychologist Risto Fried, the terrorist's interpersonal world is characterized by three categories of people: the terrorist's idealized heroes; the terrorist's enemies; and people one encounters in everyday life, whom the terrorist regards as shadow figures of no consequence.

However, Fried (1982:123) notes that some psychologists with extensive experience with some of the most dangerous terrorists "emphasize that the terrorist may be perfectly normal from a clinical point of view, that he may have a psychopathology of a different order, or that his personality may be only a minor factor in his becoming a terrorist if he was recruited into a terrorist group rather than having volunteered for one."

The Terrorist as Suicidal Fanatic

The other of the two approaches that have pre-dominated, the terrorist as fanatic, emphasizes the terrorist's rational qualities and views the terrorist as a cool, logical planning individual whose rewards are ideological and political, rather than financial. This approach takes into account that terrorists are often well educated and capable of sophisticated, albeit highly biased, rhetoric and political analysis.

Notwithstanding the religious origins of the word, the term "fanaticism" in modern usage, has broadened out of the religious context to refer to more generally held extreme beliefs. The terrorist is often labeled as a fanatic, especially in actions that lead to self-destruction. Although fanaticism is not unique to terrorism, it is, like "terrorism," a pejorative term. In psychological terms, the concept of fanaticism carries some implications of mental illness, but, Taylor (1988:97) points out, it "is not a diagnostic category in mental illness." Thus, he believes that "Commonly held assumptions about the relationship between fanaticism and mental illness...seem to be inappropriate." The fanatic often seems to view the world from a particular perspective lying at the extreme of a continuum.

Two related processes, Taylor points out, are prejudice and authoritarianism, with which fanaticism has a number of cognitive processes in common, such as an unwillingness to compromise, a disdain for other alternative views, the tendency to see things in black-and-white, a rigidity of belief, and a perception of the world that reflects a closed mind. Understanding the nature of fanaticism, he explains, requires recognizing the role of the cultural (religious and social) context. Fanaticism, in Taylor's view, may indeed "...be part of the cluster of attributes of the terrorist." However, Taylor emphasizes that the particular cultural context in which the terrorist is operating needs to be taken into account in understanding whether the term might be appropriate.

Deliberate self-destruction, when the terrorist's death is necessary in order to detonate a bomb or avoid capture, is not a common feature of terrorism in most countries, although it happens occasionally with Islamic funda-mentalist terrorists in the Middle East and Tamil terrorists in Sri Lanka and southern India. It is also a feature of North Korean terrorism. The two North Korean agents who blew up Korean Air Flight 858 on November 28, 1987, popped cyanide capsules when confronted by police investigators. Only one of the terrorists succeeded in killing himself, however.

Prior to mid-1985, there were 11 suicide attacks against international targets in the Middle East using vehicle bombs. Three well-known cases were the bombing of the U.S. Embassy in Beirut on April 18, 1983, which killed 63 people, and the separate bombings of the U.S. Marine barracks and the French military headquarters in Lebanon on October 23, 1983, which killed 241 U.S. Marines and 58 French paratroopers, respectively. The first instance, however, was the bombing of Israel's military headquarters in Tyre, in which 141 people were killed. Inspired by these suicide attacks in Lebanon and his closer ties with Iran and Hizballah, Abu Nidal launched "suicide squads" in his attacks against the Rome and Vienna airports in late

December 1985, in which an escape route was not planned.

The world leaders in terrorist suicide attacks are not the Islamic fundamentalists, but the Tamils of Sri Lanka. The LTTE's track record for suicide attacks is unrivaled. Its suicide commandos have blown up the prime ministers of two countries (India and Sri Lanka), celebrities, at least one naval battleship, and have regularly used suicide to avoid capture as well as simply a means of protest. LTTE terrorists do not dare not to carry out their irrevocable orders to use their cyanide capsules if captured.

No fewer than 35 LTTE operatives committed suicide to simply avoid being questioned by investigators in the wake of the Gandhi assassination. Attempting to be circumspect, investigators disguised themselves as doctors in order to question LTTE patients undergoing medical treatment, but, Vijay Karan (1997:46) writes about the LTTE patients, "Their reflexes indoctrinated to react even to the slightest suspicion, all of them instantly popped cyanide capsules." Two were saved only because the investigators forcibly removed the capsules from their mouths, but one investigator suffered a severe bite wound on his hand and had to be hospitalized for some time.

To Western observers, the acts of suicide terrorism by adherents of Islam and Hinduism may be attributable to fanaticism or mental illness or both. From the perspective of the Islamic movement, however, such acts of self-destruction have a cultural and religious context, the historical origins of which can be seen in the behavior of religious sects associated with the Shi'ite movement, notably the Assassins (see Glossary). Similarly, the suicide campaign of the Islamic Resistance Movement (Hamas) in the 1993-94 period involved young Palestinian terrorists, who, acting on individual initiative, attacked Israelis in crowded places, using home-made improvised weapons such as knives and axes. Such attacks were suicidal because escape was not part of the attacker's plan. These attacks were, at least in part, motivated by revenge.

According to scholars of Muslim culture, so-called suicide bombings, however, are seen by Islamists and Tamils alike as instances of martyrdom, and should be understood as such. The Arabic term used is *istishad*, a religious term meaning to give one's life in the name of Allah, as opposed to *intihar*, which refers to suicide resulting from personal distress. The latter form of suicide is not condoned in Islamic teachings.

There is a clear correlation between suicide attacks and concurrent events and developments in the Middle Eastern area. For example, suicide attacks increased in frequency after the October 1990 clashes between Israeli security forces and Muslim worshipers on Temple Mount, in the Old City of Jerusalem, in which 18 Muslims were killed. The suicide attacks carried out by Hamas in Afula and Hadera in April 1994 coincided with the talks that preceded the signing by Israel and the PLO of the Cairo agreement. They were also claimed to revenge the massacre of 39 and the wounding of 200 Muslim worshipers in a Hebron mosque by an Israeli settler on February 25, 1994.

Attacks perpetrated in Ramat-Gan and in Jerusalem in July and August 1995, respectively, coincided with the discussions concerning the conduct of elections in the Territories, which were concluded in the Oslo II agreement. The primary reason for Hamas's suicide attacks was that they exacted a heavy price in Israeli casualties. Most of the suicide attackers came from the Gaza Strip. Most were bachelors aged 18 to 25, with high school education, and some with university education. Hamas or Islamic Jihad operatives sent the attackers on their missions believing they would enter eternal Paradise.

Terrorist Group Dynamics

Unable to study terrorist group dynamics first-hand, social scientists have applied their understanding of small-group behavior to terrorist groups. Some features of terrorist groups, such as pressures toward conformity and consensus, are characteristic of all small groups. For whatever reason individuals assume the role of terrorists, their transformation into terrorists with a political or religious agenda takes places within the structure of the terrorist group. This group provides a sense of belonging, a feeling of self-importance, and a new belief system that defines the terrorist act as morally acceptable and the group's goals as of paramount importance. As Shaw (1988:366) explains:

Apparently membership in a terrorist group often provides a solution to the pressing personal needs of which the inability to achieve a desired niche in traditional society is the coup de grace. The terrorist identity offers the individual a role in society, albeit a negative one, which is commensurate with his or her prior expectations and sufficient to compensate for past losses. Group membership provides a sense of potency, an intense and close inter-personal environment, social status, potential access to wealth and a share in what may be a grandiose but noble social design. The powerful psychological forces of conversion in the group are sufficient to offset traditional social sanctions against violence....To the terrorists their acts may have the moral status of religious warfare or political liberation.

Terrorist groups are similar to religious sects or cults. They require total commitment by members; they often prohibit relations with outsiders, although this may not be the case with ethnic or separatist terrorist groups whose members are well integrated into the community; they regulate and sometimes ban sexual relations; they impose conformity; they seek cohesiveness through inter-dependence and mutual trust; and they attempt to brainwash individual members with their particular ideology.

According to Harry C. Holloway, M.D., and Ann E. Norwood, M.D. (1997:417), the joining process for taking on the beliefs, codes, and cult of the terrorist group "involves an interaction between the psychological structure of the terrorist's personality and the ideological factors, group process, structural organization of the terrorist group and cell, and the socio-cultural milieu of the group."

Citing Knutson, Ehud Sprinzak (1990:79), an American-educated Israeli political scientist, notes: "It appears that, as radicalization deepens, the collective group identity takes over much of the individual identity of the members; and, at the terrorist stage, the group identity reaches its peak." This group identity becomes of paramount importance. As Post (1990:38) explains: "Terrorists whose only sense of significance comes from being terrorists cannot be forced to give up terrorism, for to do so would be to lose their very reason for being." The terrorist group displays the characteristics of Groupthink, as described by I. Janis (1972).

Among the characteristics that Janis ascribes to groups demonstrating Groupthink are illusions of invulnerability leading to excessive optimism and excessive risk taking, presumptions of the group's morality, one-dimensional perceptions of the enemy as evil, and intolerance of challenges by a group member to shared key beliefs. Some important principles of group dynamics among legally operating groups can also be usefully applied to the analysis of terrorist group dynamics. One generally accepted principle, as demonstrated by W. Bion (1961), is that individual judgment and behavior are strongly influenced by the powerful forces of group dynamics.

Every group, according to Bion, has two opposing forces—a rare tendency to act in a fully cooperative, goal-directed, conflict-free manner to accomplish its stated purposes, and a stronger tendency to sabotage the stated goals. The latter tendency results in a group that defines itself in relation to the outside world and acts as if the only way it can survive is by fighting against or fleeing from the perceived enemy; a group that looks for direction to an omnipotent leader, to whom they subordinate their own independent judgment and act as if they do not have minds of their own; and a group that acts as if the group will bring forth a messiah who will rescue them and create a better world. Post believes that the terrorist group is the apotheosis of the sabotage tendency, regularly exhibiting all three of these symptoms.

Both structure and social origin need to be examined in any assessment of terrorist group dynamics. In Post's (1987) view, structural analysis in particular requires identification of the locus of power. In the autonomous terrorist action cell, the cell leader is within the cell, a situation that tends to promote tension. In contrast, the action cells of a terrorist group with a well-differentiated structure are organized within columns, thereby allowing policy decisions to be developed outside the cells.

Post found that group psychology provides more insights into the ways of terrorists than individual psychology does. After concluding, unconvincingly, that there is no terrorist mindset, he turned his attention to studying the family backgrounds of terrorists. He found that the group dynamics of nationalist-separatist groups and anarchic-ideological groups differ significantly. Members of nationalist-separatist groups are often known in their communities and maintain relationships with friends and family outside the terrorist group, moving into and out of the community with relative ease.

In contrast, members of anarchic-ideological groups have irrevocably severed ties with family and community and lack their support. As a result, the terrorist group is the only source of information and security, a situation that produces pressure to conform and to commit acts of terrorism.

Pressures to Conform

Peer pressure, group solidarity, and the psychology of group dynamics help to pressure an individual member to remain in the terrorist group. According to Post (1986), terrorists tend to submerge their own identities into the group, resulting in a kind of "group mind" and group moral code that requires unquestioned obedience to the group. As Crenshaw (1985) has observed, "The group, as selector and interpreter of ideology, is central." Group cohesion increases or decreases depending on the degree of outside danger facing the group.

The need to belong to a group motivates most terrorists who are followers to join a terrorist group. Behavior among terrorists is similar, in Post's analysis, because of this need by alienated individuals to belong. For the new recruit, the terrorist group becomes a substitute family, and the group's leaders become substitute parents. An implied corollary of Post's observation that a key motivation for membership in a terrorist group is the sense of belonging and the fraternity of like-minded individuals is the assumption that there must be considerable apprehension among members that the group could be disbanded. As the group comes under attack from security forces, the tendency would be for the group to become more cohesive.

A member with wavering commitment who attempts to question group decisions or ideology or to quit under outside pressure against the group would likely face very serious sanctions. Terrorist groups are known to retaliate violently against members who seek to drop out. In 1972, when half of the 30-member Rengo Sekigun (Red Army) terrorist group, which became known as the JRA, objected to the group's strategy, the dissenters, who included a pregnant woman who was thought to be "too bourgeois," were tied to stakes in the northern mountains of Japan, whipped with wires, and left to die of exposure. By most accounts, the decision to join a terrorist group or, for that matter, a terrorist cult like Aum Shinrikyo, is often an irrevocable one.

Pressures to Commit Acts of Violence

Post (1990:35) argues that "individuals become terrorists in order to join terrorist groups and commit acts of terrorism." Joining a terrorist group gives them a sense of "revolutionary heroism" and self-importance that they previously lacked as individuals. Consequently, a leader who is action-oriented is likely to have a stronger position within the group than one who advocates prudence and moderation. Thomas Strentz (1981:89) has pointed out that terrorist groups that operate against democracies often have a field commander who he calls an "opportunist," that is, an activist, usually a male, whose criminal activity predates his political involvement.

Strentz applies the psychological classification of the antisocial personality, also known as a sociopath or psychopath, to the life-style of this type of action-oriented individual. His examples of this personality type include Andreas Baader and Hans Joachim Klein of the Baader-Meinhof Gang and Akira Nihei of the JRA. Although the opportunist is not mentally ill, Strentz explains, he "is oblivious to the needs of others and unencumbered by the capacity to feel guilt or empathy."

By most accounts, Baader was unpleasant, constantly abusive toward other members of the group, ill-read, and an action-oriented individual with a criminal past. Often recruited by the group's leader, the opportunist may eventually seek to take over the group, giving rise to increasing tensions between him and the leader. Often the leader will manipulate the opportunist by allowing him the fantasy of leading the group.

On the basis of his observation of underground resistance groups during World War II, J.K. Zawodny (1978) concluded that the primary determinant of underground group decision making is not the external reality but the psychological climate within the group. For action-oriented terrorists, inaction is extremely stressful. For action-oriented members, if the group is not taking action then there is no justification for the group. Action relieves stress by reaffirming to these members that they have a purpose. Thus, in Zawodny's analysis, a terrorist group needs to commit acts of terrorism in order to justify its existence.

Other terrorists may feel that their personal honor depends on the degree of violence that they carry out against the enemy. In 1970 Black September's Salah Khalef ("Abu Iyad") was captured by the Jordanians and then released after he appealed to his comrades to stop fighting and to lay down their arms. Dobson (1975:52) reports that, according to the Jordanians, Abu Iyad "was subjected to such ridicule by the guerrillas who had fought on that he reacted by turning from moderation to the utmost violence."

Pearlstein points out that other examples of the political terrorist's self-justification of his or her terrorist actions include the terrorist's taking credit for a given terrorist act and forewarning of terrorist acts to come. By taking credit for an act of terrorism, the terrorist or terrorist group not only advertises the group's cause but also communicates a rhetorical self-justification of the terrorist act and the cause for which it was perpetrated. By threatening future terrorism, the terrorist or terrorist group in effect absolves itself of responsibility for any casualties that may result.

Terrorist Rationalization of Violence

Living underground, terrorists gradually become divorced from reality, engaging in what Ferracuti (1982) has described as a "fantasy war." The stresses that accompany their underground, covert lives as terrorists may also have adverse social and psychological consequences for them. Thus, as Taylor (1988:93) points out, although "mental illness may not be a particularly helpful way of conceptualizing terrorism, the acts of terrorism and membership in a terrorist organization may well have implications for the terrorist's mental health."

Albert Bandura (1990) has described four techniques of moral disengagement that a terrorist group can use to insulate itself from the human consequences of its actions. First, by using moral justification terrorists may imagine themselves as the saviors of a constituency threatened by a great evil. For example, Donatella della Porta (1992:286), who interviewed members of left-wing militant groups in Italy and Germany, observed that the militants "began to perceive themselves as members of a heroic community of generous people fighting a war against 'evil.'"

Second, through the technique of displacement of responsibility onto the leader or other members of the group, terrorists portray themselves as functionaries who are merely following their leader's orders. Conversely, the terrorist may blame other members of the group. Groups that are organized into cells and columns may be more capable of carrying out ruthless operations because of the potential for displacement of responsibility. Della Porta's interviews with left-wing militants suggest that the more compartmentalized a group is the more it begins to lose touch with reality, including the actual impact of its own actions. Other manifestations of this displacement technique include accusations made by Asahara, the leader of Aum Shinrikyo, that the Central Intelligence Agency (CIA) used chemical agents against him and the Japanese population.

A third technique is to minimize or ignore the actual suffering of the victims. As Bonnie Cordes (1987) points out, terrorists are able to insulate themselves from moral anxieties provoked by the results of their hit-and-run attacks, such as the use of time bombs, by usually not having to witness first-hand the carnage resulting from them, and by concerning themselves with the reactions of the authorities rather than with civilian casualties. Nevertheless, she notes that "Debates over the justification of violence, the types of targets, and the issue of indiscriminate versus discriminate killing are endemic to a terrorist group." Often, these internal debates result in schisms.

The fourth technique of moral disengagement described by Bandura is to dehumanize victims or, in the case of Islamist groups, to refer to them as "the infidel." Italian and German militants justified violence by depersonalizing their victims as "tools of the system," "pigs," or "watch dogs." Psychologist Frederick Hacker (1996:162) points out that terrorists transform their victims into mere objects, for "terroristic thinking and practices reduce individuals to the status of puppets." Cordes, too, notes the role reversal played by terrorists in characterizing the enemy as the conspirator and oppressor and accusing it of state terrorism, while referring to themselves as "freedom fighters" or "revolutionaries." As Cordes explains, "Renaming themselves, their actions, their victims and their enemies accords the terrorist respectability."

By using semantics to rationalize their terrorist violence, however, terrorists may create their own self-destructive psychological tensions. As David C. Rapoport (1971:42) explains:

All terrorists must deny the relevance of guilt and innocence, but in doing so they create an unbearable tension in their own souls, for they are in effect saying that a person is not a person. It is no accident that left-wing terrorists constantly speak of a "pig-society," by convincing *themselves* that they are confronting animals they hope to stay the remorse which the slaughter of the innocent necessarily generates.

Expanding on this rationalization of guilt, D. Guttman (1979:525) argues that "The terrorist asserts that he loves only the socially redeeming qualities of his murderous act, not the act itself." By this logic, the conscience of the terrorist is turned against those who oppose his violent ways, not against himself. Thus, in Guttman's analysis, the terrorist has projected his guilt outward. In order to absolve his own guilt, the terrorist must claim that under the circumstances he has no choice but to do what he must do. Although other options actually are open to the terrorist, Guttman believes that the liberal audience legitimizes the terrorist by accepting this rationalization of murder.

Some terrorists, however, have been trained or brainwashed enough not to feel any remorse, until confronted with the consequences of their actions. When journalist Eileen MacDonald asked a female ETA commando, "Amaia," how she felt when she heard that her bombs had been successful, she replied, after first denying being responsible for killing anyone.

"Satisfaction the bastards, they deserved it. Yes, I planted bombs that killed people." However, MacDonald felt that Amaia, who had joined the military wing at age 18, had never before questioned the consequences of her actions, and MacDonald's intuition was confirmed as Amaia's mood shifted from bravado to despondency, as she buried her head in her arms, and then groaned: "Oh, God, this is getting hard," and lamented that she had not prepared herself for the interview.

When Kim Hyun Hee (1993:104), the bomber of Korean Air Flight 858, activated the bomb, she had no moral qualms. "At that moment," she writes, "I felt no guilt or remorse at what I was doing; I thought only of completing the mission and not letting my country down." It was not until her 1988 trial, which resulted in a death sentence — she was pardoned a year later because she had been brainwashed — that she felt any remorse. "But being made to confront the victims' grieving families here in this courtroom," she writes, "I finally began to feel, deep down, the sheer horror of the atrocity I'd committed." One related characteristic of Kim, as told by one of her South Korean minders to McDonald, is that she had not shown any emotion whatsoever to anyone in the two years she (the minder) had known her.

The Terrorist's Ideological or Religious Perception

Terrorists do not perceive the world as members of governments or civil society do. Their belief systems help to determine their strategies and how they react to government policies. As Martha Crenshaw (1988:12) has observed, "The actions of terrorist organizations are based on a subjective interpretation of the world rather than objective reality."The variables from which their belief systems are formed include their political and social environments, cultural traditions, and the internal dynamics of their clandestine groups. Their convictions may seem irrational or delusional to society in general, but the terrorists may nevertheless act rationally in their commitment to acting on their convictions.

According to cognitive theory, an individual's mental activities (perception, memory, and reasoning) are important determinants of behavior. Cognition is an important concept in psychology, for it is the general process by which individuals come to know about and make sense of the world. Terrorists view the world within the narrow lens of their own ideology, whether it be Marxism-Leninism, anarchism, nationalism, Islamic fundamentalism, or some other ideology. Most researchers agree that terrorists generally do not regard themselves as terrorists but rather as soldiers, liberators, martyrs, and legitimate fighters for noble social causes.

Those terrorists who recognize that their actions are terroristic are so committed to their cause that they do not really care how they are viewed in the outside world. Others may be just as committed, but loathe to be identified as terrorists as opposed to freedom fighters or national liberators. Kristen Renwick Monroe and Lina Haddad

Kreidie (1997) have found *perspective*—the idea that we all have a view of the world, a view of ourselves, a view of others, and a view of ourselves in relation to others—to be a very useful tool in understanding fundamentalism, for example.

Their underlying hypothesis is that the perspectives of fundamentalists resemble one another and that they differ in significant and consistent ways from the perspectives of non fundamentalists. Monroe and Kreidie conclude that "fundamentalists see themselves not as individuals but rather as symbols of Islam." They argue that it is a mistake for Western policymakers to treat Islamic fundamentalists as rational actors and dismiss them as irrational when they do not act as predicted by traditional cost/benefit models. "Islamic fundamentalism should not be dealt with simply as another set of political values that can be compromised or negotiated, or even as a system of beliefs or ideology—such as socialism or communism—in which traditional liberal democratic modes of political discourse and interaction are recognized."

They point out that "Islamic fundamentalism taps into a quite different political consciousness, one in which religious identity sets and determines the range of options open to the fundamentalist. It extends to all areas of life and respects no separation between the private and the political."

Existing works that attempt to explain religious fundamentalism often rely on modernization theory and point to a crisis of identity, explaining religious fundamentalism as an antidote to the dislocations resulting from rapid change, or modernization. Islamic fundamentalism in particular is often explained as a defense against threats posed by modernization to a religious group's traditional identity. Rejecting the idea of fundamentalism as pathology, rational choice theorists point to unequal socioeconomic development as the basic reason for the discontent and alienation these individuals experience.

Caught between an Islamic culture that provides moral values and spiritual satisfaction and a modernizing Western culture that provides access to material improvement, many Muslims find an answer to resulting anxiety, alienation, and disorientation through an absolute dedication to an Islamic way of life. Accordingly, the Islamic fundamentalist is commonly depicted as an acutely alienated individual, with dogmatic and rigid beliefs and an inferiority complex, and as idealistic and devoted to an austere lifestyle filled with struggle and sacrifice.

In the 1990s, however, empirical studies of Islamic groups have questioned this view. V. J. Hoffman-Ladd, for example, suggests that fundamentalists are not necessarily ignorant and downtrodden, according to the stereotype, but frequently students and university graduates in the physical sciences, although often students with rural or traditionally religious backgrounds.

In his view, fundamentalism is more of a revolt of young people caught between a traditional past and a secular Western education. R. Euben and Bernard Lewis argue separately that there is a cognitive collision between Western and fundamentalist worldviews. Focusing on Sunni fundamentalists, Euben argues that their goals are perceived not as self-interests but rather as moral imperatives, and that their worldviews differ in critical ways from Western worldviews.

By having moral imperatives as their goals, the fundamentalist groups perceive the world through the distorting lens of their religious beliefs. Although the perceptions of the secular Arab terrorist groups are not so clouded by religious beliefs, these groups have their own ideological imperatives that distort their ability to see the world with a reasonable amount of objectivity. As a result, their perception of the world is as distorted as that of the fundamentalists. Consequently, the secular groups are just as likely to misjudge political, economic, and social realities as are the fundamentalist groups.

For example, Harold M. Cubert argues that the Popular Front for the Liberation of Palestine (PFLP), guided by Marxist economic ideology, has misjudged the reasons for popular hostility in the Middle East against the West, "for such hostility, where it exists, is generally in response to the threat which Western culture is said to pose to Islamic values in the region rather than the alleged economic exploitation of the region's inhabitants." This trend has made the PFLP's appeals for class warfare irrelevant, whereas calls by Islamist groups for preserving the region's cultural and religious identity have been well received, at least among the non secular sectors of the population.

4

Terrorism and the Power of Language

News Attempt to Capture the World in Language

It comprises a social frame according to which people understand and make sense of their experience. This consensual power we have granted words, at the same time it allows us to interact with our context and with others, inherently and explicitly dictates a normative structure that orientates the way we live. Words convey a specific meaning that is derived contextually and socially. News as discourse, as a product of language, shapes our perception of the world. It too is embedded in a specific paradigm of values. But it can either conform to it and reaffirm a given representation of reality, or challenge it and potentially modify our appreciation. Journalism has that capacity; it can impact our truths, our claims of knowledge and our notions of justice. Journalists can be an authentic catalyst for change but they can also legitimize reductionism, stagnation and stereotypes.

Terrorism as Core Issue

It is a global preoccupation, and the terminology that has become inextricable from the event itself ("terrorist," "war on terror," "weapons of mass destruction") resonates imperatively throughout the globe. But words have a variety of definitions; they should not be condemned to have one fixed, loaded meaning forever. John Hartley (1982) states that news discourse plays an important role in assigning one single meaning (uni-accentual value) to signs. One of its tasks is to ensure that one particular meaning is preferred to another.

Mainly, news has taken on one rhetoric to refer to the phenomena of terrorism, its perpetrators and its deeds. The media have devoted their attention to the eventfulness of the conflict, to its dramatic nature, and to the distinction between "us" and "them." They have rarely stopped to report about the origins, the motivations and the facts surrounding a deeply rooted symptomatic social, political and economical conflict. An unbalanced, emotive depiction of a matter as sensitive and complex as terrorism, has inflated a conflict that awaits urgent solutions.

The traditional perspective of reporting terrorism implemented in most democratic systems does not favor terrorist activities through media in the way it has been said to do. Within the more traditional way of reporting terrorist events there seems to be a general misconception that:

1) Terrorist action is an overall effective method for reaching recognition, therefore it is successful.
2) Terrorism can't survive without the media. In fact, terrorists use media to their full advantage and find legitimacy in them.

For one, even though the coverage of terrorism can get them recognition, what terrorists also strive for is social acceptance and sympathy. The specific way in which media portray them does not lead to either one of these.

Still, the argument held by those in power that media could legitimize actions of terrorists, accounts for the choice of various media outlets to depict terrorism using strongly loaded criminal terms:

- Groups defined as terrorists are referred to as "murderers," "thugs" and "assassins." They are "shadowy."
- Headlines label specific actions as terrorism by including the words "terror," and "terrorism," contrasting them with headlines that use terms such as "freedom fighter."
- The actions and objectives of terrorists are labeled as "mad" or "as a game."

An event orientation of the media focuses on the sensational aspects of terrorism: on the blood, the guts and the drama. The motivations or otherwise sensitive issues around terrorism do not make it into most news coverage. The political rationale behind this is that government's fear sympathetic reporting of any sort could destabilize their efforts to combat terrorism. In regard to the arguments presented before, it is fair to say that in almost the majority of cases, terrorists will not achieve their most important objective: legitimacy.

The alternative perspective

An alternative to the more orthodox way of covering terrorism posits that:

- Official semantic framings of terrorism should not be reproduced indiscriminately.
- State terrorism as well as right wing terrorism should be covered.
- The coverage of journalism should be more analytical and explanatory of the phenomenon of terrorism.

EXAMPLES:

- The case of Vrye Weekblad and Weekly Mail in South Africa in the eighties is an example of how alternative coverage can expose state terrorism in the benefit of society.

- *An Phoblacht*, the newspaper of the Sinn Fein party in Northern Ireland, attempted to explain the social and political issues behind the conflict in the region.
- Peace Reporting.
- Responsible Reporting.

Facing the New Terrorism

According to Secretary of State Madeleine K. Albright, last summer's embassy bombings in Kenya and Tanzania were not two more examples of old-fashioned terrorism. "What is new," she declared, "is the emergence of terrorist coalitions that do not answer fully to any government, that operate across national borders and have access to advanced technology." The bomb victims, she claimed, were caught up "in a new kind of confrontation that looms as a new century is about to begin ... a clash between civilization itself and anarchy — between the rule of law and no rules at all."

The secretary's words would have been accurate had they been uttered a century earlier, when a loose-knit transnational movement quite literally devoted to the promotion of anarchy wreaked havoc across the globe. From 1894 to 1901, anarchists managed to assassinate the president of France, the prime minister of Spain, the empress of Austria, the king of Italy, and the president of the United States. All this was accomplished without downloading weapons diagrams from the Internet; they relied instead on manuals such as Johann Most's widely distributed pamphlet, The Science of Revolutionary Warfare: A Little Handbook of Instruction in the Use and Preparation of Nitroglycerin, Dynamite, Gun-Cotton, Fulminating Mercury, Bombs, Fuses, Poisons, etc.

Some anarchists showed no scruples in inflicting large numbers of civilian casualties. As the bomber of a crowded Parisian cafe put it at his trial, "[Anarchists] do not spare bourgeois women and children, because the wives and children of those they love are not spared either." And authorities responded then as they do today; one British police officer wrote in 1898, "Murderous organizations have increased in size and scope; they are more daring, they are served by the more terrible weapons offered by modern science, and the world is nowadays threatened by new forces which ... may someday wreak universal destruction."

Extralegal political violence by individuals and groups has occurred throughout history, hysterical media coverage of today's terrorism "crises" notwithstanding. Even religious terrorism is nothing new. "Thug," "zealot," and "assassin" are now generic terms of abuse, but each entered the language as the name of a religious terrorist movement centuries ago (emerging from Hinduism, Judaism, and Islam, respectively). And contrary to official statements about the grave danger terrorism poses, most American national security experts and bureaucracies have traditionally paid it scant attention. Terrorism kills fewer Americans than does lightning, they say in private — which happens to be true — and overreaction to it is therefore a sucker's move. Conventional state violence is a far more serious and pressing threat, they insist.

In the last few years, however, this dismissive attitude has come under fire. Some argue that terrorism merits a higher priority now because other threats, like great-power war, have grown so remote. Others say the danger is mounting in absolute as well as relative terms because of changes in terrorist motivations, methods, and organization. And almost everyone was scared by the 1995 nerve gas attack on a Tokyo subway by the Japanese cult Aum Shinrikyo. As terrorism expert Brian Jenkins has remarked, before that incident the debate over terrorists using weapons of mass destruction (WMD) pitted "disbelievers," who thought it would not happen because it had not, against "Murphy's Lawyers," who thought catastrophes were inevitable. The Aum Shinrikyo attack proved the first camp wrong but not the second camp right.

The appearance of several good new books on the subject is therefore a welcome surprise — welcome because they help answer crucial questions and surprising because, with a few exceptions, the literature on terrorism thus far has not been especially distinguished. Bruce Hoffman's Inside Terrorism is a concise yet authoritative survey of trends in terrorism past and present. Philip Heymann's Terrorism and America gives us a sensible guide to how the U.S. government should respond. And three other volumes — America's Achilles' Heel, by Richard Falkenrath, Robert Newman, and Bradley Thayer; The Ultimate Terrorists, by Jessica Stern; and Terrorism with Chemical and Biological Weapons, edited by Brad Roberts — focus specifically on the WMD threat. All five books combine serious scholarship with practical wisdom, and the volumes by Hoffman and Falkenrath et al., are particularly comprehensive. What is most interesting, however, is that these independent efforts display a remarkable degree of consensus on the nature of contemporary challenges and what should be done about them.

Discussion of the future threat, the authors agree, should begin with recognition that previous predictions of rampant catastrophic terrorism have proved grossly exaggerated. Although WMD have been available for decades, terrorists have generally not tried to acquire them, let alone use them against actual targets. Until recently, in fact, terrorists had not even begun to exploit the full destructive potential of conventional weapons. The reason is simple: most terrorism involves carefully calibrated acts of symbolic violence designed to advance a political, social, or bureaucratic agenda, and true mass murder could be counterproductive. It might stigmatize the cause in the eyes of important international or domestic audiences, provoke massive retaliation from the authorities, or spark conflict within the terrorist group itself. Garden-variety terrorists, as Jenkins once famously put it, want a lot of people watching, not a lot of people dead.

To the extent that such rational calculations are indeed the reason we have escaped catastrophic terrorism in the past, they should continue to impose restraints in the future on mainstream terrorist groups, those with a known address and somewhat limited objectives. Hoffman's book traces the history of such organizations from the anarchist and leftist terrorism of the late nineteenth century, through the nationalist and separatist terrorism of the colonial and postcolonial era, to the international and state-sponsored terrorism of the 1960s, 1970s, and 1980s. In recent years, he notes, religious terrorism has increasingly come to the fore. And while the number of terrorist attacks has declined, the number of casualties per attack has increased.

This is where the story starts to get truly worrisome. All five books warn of the increasing danger of non-traditional terrorists, whose behavior is less predictable and who might seek to maximize bloodshed. Three such types quickly come to mind: religious fanatics who consider violence a sacramental act or believe they are the direct instruments of divine retribution; eschatological cults with a penchant for violence; and disturbed or hate-filled activists who want to inflict pain on a grand scale. The good news is that such groups and individuals are few and far between. The bad news is that all three types do exist — as the World Trade Center bombing, the Japanese subway attack, and the Oklahoma City bombing testify — and there is reason to believe their numbers may be growing. (Establishing whether this is indeed the case, and if so why, should be a top priority for future research.)

More bad news is the fact that WMD capability is gradually coming within range of many sub state actors through the general diffusion of scientific skills and dual-use technologies. To be sure, terrorists content to cause dozens or hundreds of casualties will probably stick with conventional methods. Yet the more ambitious of them might be tempted by chemical, biological, or nuclear alternatives, whose respective profiles are summarized neatly by Falkenrath et al.:

Chemical weapons suitable for mass casualty attacks can be acquired by virtually any state and by nonstate actors with moderate technical skills. Certain very deadly chemical warfare agents can quite literally be manufactured in a kitchen or basement in quantities sufficient for mass-casualty attacks.

Many states and moderately sophisticated non state actors could construct improvised but effective biological weapons. . . . Culturing the required microorganisms, or growing and purifying toxins, is inexpensive and could be accomplished by individuals with college level training in biology and a basic knowledge of laboratory technique. Acquiring the seed stocks for pathogenic microorganisms is also not particularly difficult.

Nuclear weapons are within the reach of tens of states, with the most significant constraint being the ability to produce plutonium or highly enriched uranium. If this obstacle were avoided through the theft or purchase of fissile material, almost any state with a reasonable technical and industrial infrastructure could fabricate a crude nuclear weapon, . . . [as could] some exceptionally capable non state actors.

"Weapons of all three classes," they add, "are deliverable against a wide range of targets, and defense is difficult." There seems to be general agreement that of the three, chemical use is the most likely scenario, nuclear use the least likely, and biological use the scariest because it is both relatively easy and highly deadly.

The intersection of the trend involving motive and the trend involving opportunity is what made the Aum

Shinrikyo case so disturbing. After all, as Stern points out, this was a group whose members penned ditties such as the following:

It came from Nazi Germany, a dangerous chemical weapon,

Sarin, Sarin!

If you inhale the mysterious vapor, you will fall with bloody vomit from your mouth,

Sarin, sarin, sarin, the chemical weapon!

Song of Sarin the Brave. . . .

Given these trends and the fact that terrorists tend to copy each other's methods, one can understand why all these books consider catastrophic terrorism a seismic event whose probability is low but rising. Weapons of mass destruction will not become the car bombs of the next few decades, they argue, but the situation nevertheless calls for precautionary measures to reduce vulnerability, head off attacks, and manage potential consequences.

Heymann offers a range of intelligent, if unsurprising, suggestions for handling traditional terrorist threats, from avoiding concessions, cooperating with allies, and prosecuting suspects, to relying on crack hostage-rescue teams or military retaliation where appropriate. He makes a persuasive case for avoiding overreaction, arguing both that it would be tragic for democracies to abandon their cherished freedoms and principles in a quest for absolute security and that there is little reason to believe that a heavy-handed approach to counterterrorism would work. After much hemming and hawing, however, even he seems to favor some increase in domestic intelligence gathering, concluding that "the limited threat to uninhibited discussion posed by even reasonable efforts to monitor organizations preaching violence is a price worth paying to prevent political violence."

The other authors agree that in addition to the standard responses, governments can and should do more to prepare for the worst-case WMD scenarios. They make a number of strikingly similar policy recommendations, among which are the following:

Improve intelligence collection and analysis. The groups that cause the greatest concern — religious fanatics, cults, and freelance extremists — are precisely those that usually fly below the radar screen of standard intelligence collection. They tend to be isolated from society and may have no prior criminal history. The U.S. government should therefore change its practices in order to track the activities of such groups wherever they might be and monitor indicators of small-scale WMD programs. It also needs to improve epidemiological surveillance systems at home and abroad. These could alert authorities to accidents from covert biological weapons programs and provide early detection of attacks in progress, which could be crucial for effective response.

Restructure bureaucratic organizations. Current U.S. efforts to prevent and respond to WMD terrorism are spread across a vast number of agencies at different levels of government with little real coordination or direction. Bureaucratic styles and missions clash; information is compartmentalized and left unanalyzed; some tasks are duplicated while others slip through the cracks. Each author has a different pet solution to this problem, but two ideas appear frequently — that the government should set up a central interagency catastrophic terrorism response center and that the Department of Defense should play a greater role in handling WMD attacks on U.S. soil. Many experts feel that weapons of mass destruction are simply not a specialty of the FBI, and where the potential for catastrophic terrorism is concerned, the FBI's reactive law enforcement approach needs to be supplemented by — if not subordinated to — the more aggressive national security approach of the White House and the Pentagon.

Enhance domestic preparedness. Quick and appropriate responses could drastically limit the scale of disaster should a catastrophic terrorist attack ever occur, while delay or chaos would multiply the mayhem. All the authors agree, therefore, that more needs to be done to protect critical infrastructures; coordinate federal, state, and local readiness; train "first responders"; stockpile medicines and vaccines; and develop and disseminate technologies for identifying WMD use. Here the 1996 Defense Against Weapons of Mass Destruction Act — the Nunn-Lugar-Domenici amendment — represents a good start but needs regular funding at much higher levels and should be followed up with further efforts.

Deal with the Russian challenge. The single best way to lower the probability of nuclear terrorism is to control fissile material, and vast stockpiles of such material lie poorly protected in Russia. Helping the Russians control these and other material and human remnants of their massive WMD programs is critical. Here again, the Nunn-Lugar-Domenici plan represents a good start, but it needs vastly greater funding and complementary measures to boot.

Outlaw WMD possession by sub state actors. To deter future terrorists and lay the groundwork for prosecuting others, several authors suggest, the U.S. government should support efforts to make WMD possession by individuals or sub state groups a universal crime under international law. Furthermore, the government should educate relevant industries about the WMD terrorist threat and push them to adopt strict self-policing and reporting measures, so that covert or unauthorized WMD programs can be identified and stopped early.

To the credit of the Clinton administration and some in Congress, in recent years the government has begun to move in the right direction in many of these areas. The president's recent proposals to increase preparedness against chemical and biological terrorism are sound and deserve support, as do his calls for more funding for efforts to manage Russian WMD materiel and scientists. But ships of state take a long time to turn and — rhetoric aside — it will be a while before the new course is reached. What these books show clearly is that the main obstacles ahead are not intellectual or practical but rather political and bureaucratic.

Implementing the agenda outlined above, for example, would take little more than another couple of billion dollars a year, combined with strong leadership and a coherent national strategy, and would help prepare the nation against covert WMD use by rogue states as well as terrorists. Yet worthy measures such as improving epidemiological surveillance or controlling foreign fissile materials attract few powerful backers because they are unusual, unsexy, and provide comparatively few opportunities for pork. Grandiose schemes like ballistic missile defense, in contrast, enjoy funding and attention far out of proportion to their true practical value. (Stern voices a common complaint when she writes that "ballistic missiles are the least likely method of [WMD] delivery, and yet American taxpayers spend roughly ten times as much on defense against [them] as on the entire program to prevent WMD terrorism.")

In the end, the answer should be neither complacency nor hysteria but rather modest, sustained investment in countermeasures and preparedness. Individuals take out insurance policies all the time to hedge against disasters that will probably never occur. This is one case where governments should do the same — and count themselves lucky if the premiums are ultimately wasted.

The Solution to Terrorism

The case for shifting power and authority from the national government to the states is usually made on philosophical or historical grounds. While there is clearly a case to be made for classical federalism (strong states in tandem with a strong but distinctly limited national government), little has been said about the efficiency and suitability of the classical federal model as a response to contemporary world threats.

The federal government plans to increase Homeland Defense spending by 10% in FY 2005, including $33.8 billion in discretionary budget authority. It's a value more than 50% over the net worth of Paul Allen, the third-richest person in America, and is easily matched or bettered by private-sector spending on anti-terrorism measures. Terrorism has forced both the private and public sectors to evaluate how they distribute resources. Terrorists themselves understand that terrorism can leverage vastly greater damage on its victims than it costs to carry out.

Targets can't afford to ignore the terrorist threat altogether, but much of what has been done in the United States in response to terrorism has been symptomatic rather than systematic. The result is that actual security has only been increased somewhat, while the incentive structure that makes terrorism attractive to insurgents and enemies of the United States has been left virtually untouched. The likely consequence is that a large volume of resources will be needlessly expended in pursuit of an imperfect sense of security.

It is surprising that there hasn't been a broader debate on more economically efficient means of defending the United States against terrorism. It would be logically absurd to try to spend our way into security, since we could in fact devote the entirety of our gross domestic product on security measures without actually guaranteeing that no further attacks would take place. Yet it has so far eluded us in our response to terrorism to consider changing systems and structures in order to make them more robust in response to the threat of attack.

With the ever-increasing scope and reach of the national government has come an increasing concentration of power and control at the national level. The same effect has simultaneously played out in the private sector as more challenging tax and regulatory structures have enhanced the environment for large businesses at the expense of a more vibrant market for small- and medium-sized businesses, since those smaller firms are less able to effectively shield themselves from those taxes and regulations through accountancy, legal action, and political favoritism. The net effect of these trends has been to concentrate power and control, both geographically and structurally.

Today, one national government in Washington, DC, commands the powers of the vastly greater part of what was intended as a balanced federal system. National authority extends far beyond simply the reach of national-level agencies. Many (if not most) state and local governments depend upon direct fiscal support from the national government in order to carry out even their most basic activities, like building infrastructure and maintaining school systems. The net impact has been to create a system in which local and state governments are neutered and have become largely incapable of self-maintenance.

Again, the same effect is widespread in the private sector as well, where fewer businesses are independent and large organizations concentrate decision-making authority at the top echelons.

This is precisely the opposite of a healthy structure for sustainability and robustness in response to terrorism. The attacks of September 11th were brutally effective at causing financial and economic hardship precisely because they brought business activity in Manhattan to a standstill. With so much of the nation's business and financial activity dependent upon that one geographic space, a single disaster wrought enormous distress.

Similarly, with so much of the government's power and authority concentrated within Washington, DC, it too is both a geographic and structural linchpin. Discussions of "government continuity" in the event of terror attacks only serves to confirm the observation that the system has been rendered highly unstable by the perpetual efforts to push government responsibilities up the food chain.

Heavy concentration of power and authority in a relatively small number of agency directors and officials, many of whom live in or around Washington, DC.
Power and responsibility, wherever reasonably possible, should be pushed back "down" to the states. The sheer difficulty of trying to similarly disrupt operations for each of 50 state governments, plus the national government itself, makes it nearly impossible to conceive of any attack or series of attacks that could effectively disrupt the self-government of the entire United States.

Intelligence-Gathering

The CIA and FBI have failed on several high-profile occasions to effectively gather the necessary evidence and intelligence needed to thwart terrorist attacks. Much of the best footwork on terror issues has been done at the local level, but many state and local law-enforcement agencies remain dependent upon federal funding and training. A healthier approach would enhance the independence of local authorities to coordinate themselves independent of federal agencies, especially in case of damage to the federal agencies' ability to respond to attacks.

Budgets and Fiscal Authority

The national government holds sway over vast reaches of local budgetary authority, from education to infrastructure-building. States and local agencies need more responsibility for and ability to handle their own financial conditions in order to reduce the risk that catastrophic events on a national scale will cause those local agencies to fall apart as well.

Business Continuity

A vast number of major companies are concentrated in a handful of states, including New York and California, while states like Iowa and Indiana and South Dakota have few or none. Businesses would be wise to consider spreading out their risk pool geographically in order to reduce their exposure in any one location.

Transportation Bottlenecks

The nation's transportation system, particularly for air travel, depends upon a small number of hub airports. Disasters or delays caused at any one of these airports can cause a ripple effect of trouble throughout the system. The sustainability of regular air travel in case of a major terrorist disruption at O'Hare, LAX, John F. Kennedy, Dallas-Fort Worth, or Atlanta would be difficult at best.

Southwest Airlines has shown that the hub-and-spoke model isn't the only way to run an airline. Given the risk that a catastrophic event could be targeted at one of these airports, it's difficult to understand how the major airlines continue to justify concentrating huge shares of their equipment and labor at those few destinations. A major event at O'Hare could certainly spell the end for United, just as one at Dallas-Fort Worth could mean the same for American.

Business-Sector Resilience

Complicated tax and regulatory conditions tend to favor large firms, since they have the resources to afford the accountancy and legal advice that can help them avoid the impact of those taxes and regulations. Further, large firms are more capable than small firms of buying political influence through campaign contributions and lobbying. The resulting market conditions favor business concentration, which exposes the economy to risks similar to those of government concentration.

An environment that offers small firms a level playing field in terms of taxation and regulation would increase the robustness of the private sector by increasing the number of sustainable firms that would be likely to survive major damaging events. Small firms are particularly sensitive to challenging tax and regulatory climates, since they are forced to spend a disproportionate share of their resources on compliance, which reduces their ability to invest in growth and development.

5

Methods of Terrorism and the Balance between Freedom and Security

A New Quality of Terrorism

"The world after September 11 presents a particular challenge to all of us." (A. Lewis, 2002: 61) – this or something like this is the tenor of just about all articles in journals or newspapers that are dealing with the terrorist attacks of September 11, 2001. In an address to his cabinet, U.S. President Bush called the events a "deadly and deliberate attack by a different enemy than we have ever faced". The German Federal Minster of the Interior Otto Schily emphasized that with New York – as "a symbol for the desire for freedom in this world, for democracy in this world" – being the target of the terrorist strike and more than 80 nations having citizens among the victims, not only the USA were attacked, but freedom, democracy and the consciousness of the Western world in general.

And indeed, September 11, 2001, ultimately made clear to the Western world that they were confronted with the development of a new quality of terrorism. A great deal of the knowledge one had about terrorist organizations is now obsolete. "Even distinctions such as international and domestic terrorism are decreasingly meaningful because technological advances and corporate globalization facilitate more complex and flexible ways to organize terrorist activities" (Turk, 2004: 276). The new terrorist threat was perceived as de-regionalized and de-individualized – it was no longer possible to identify specific, known perpetrators and their limited environment, but the danger arose from "impersonal networks and organizations harbored in the diffusion of Islamic fundamentalism".

In one sentence: the motives and organization of terrorism became less simple and local in recent years. Now the greatest threat no longer emanates from nationalist or material, but from ideological, especially religious concerns. The purpose of this paper is to discuss, how the reaction to this new quality of terrorism affects the balance between freedom and civil liberties on the one side, and national and public security on the other side. On the example of the Federal Republic of Germany it is examined, what civil liberties are affected and how. Before doing this, the issue of the perception of terrorism is shortly illuminated. The last part of the paper calls for the engagement of social scientists in response to the curbing of civil liberties and finally gives an out-look on the future development of the balance between freedom and security.

Terrorism as Social Construction

Notwithstanding the points mentioned in the beginning, terrorism not only has to be recognized as a fact or an event, but as a "social construction". This is especially important when talking about reactions to terrorist attacks or fear of and reactions to the threat of terrorism. "Terrorism is not a given in the real world, but is instead an interpretation of events and their presumed causes. And these interpretations are not unbiased attempts to depict truth but rather efforts to manipulate perceptions to promote certain interests at the expense of others".

The so-called "labeling theory" or "societal reaction theory" says that labels given by society define the incident and determine the reaction to this incident. Some sociologists take the stand that the individuals and groups of a society do in fact not at all react to the actual event, but to labels that are assigned to this event. In the case of an emergency, the reaction of society therefore depends on how the occurrence is defined or labeled. "Frequently, it is the perceived threatening nature of the event at the time of the occurrence rather than the actual enduring consequences of the event that instigates the official and social responses".

What implications has this theory for the response to terrorism? The attacks of September 11, 2001, have been publicly perceived as an emergency. When such an emergency exists, people are likely to "rally around the flag" to support the country or the president and accept whatever measures are taken to respond to the emergency. In bringing the public to the point where they see current events as constituting an emergency, the media is playing an important role. In the case of the attack on the World Trade Center in New York, the media coverage was extensive: the whole world immediately was made aware of the events. The event was labeled as an "attack worse than Pearl Harbor", it was defined as "the bloodiest day on American soil since our Civil War" and titled "a clear and present danger" and an "act of war". In terms of societal reaction, the taking of severe measures was predictable.

The Balance between Security and Freedom

As the American Chief Justice William Rehnquist suggests, "national emergencies shift the balance between freedom and order toward order – in favor of the government's ability to deal with the conditions that threaten the national well-being". It is largely accepted that there is a trade-off between security and civil liberties – in general the one can only be enlarged at the expense of the other.

According to the preamble of the U.S. Constitution, the fundamental purposes of the federal government are to "establish justice, insure domestic tranquility, provide for the common defense, promote the general welfare, and secure the blessings of liberty ...".

But how the trade-offs between the two values order and liberty, which the Constitution is dedicated to, should be made and when they are acceptable, is not explained. The competition between those two values is also prominently articulated in the famous quote by the former U.S. President Benjamin Franklin: "They that can give up essential liberty to obtain a little temporary safety deserve neither liberty nor safety". As expressed by this quote, often the measures taken to improve security in the name of liberty, are in fact eroding the freedoms being defended.

The terrorist acts of Sept. 11 riveted the attention of the nation as hijacked commercial airliners hit first one tower and then the second tower of the World Trade Center complex in New York. Given the magnitude of the attacks and the location in one of the country's largest media centers, live television and radio coverage was almost immediate. CBS aired a live shot of the second plane crashing into the tower. All of the networks ran live coverage of the towers disintegrating and collapsing. News of the attacks monopolized all channels of communication throughout the day. Several cable channels preempted their programming in favor of CNN coverage or suspended broadcasting entirely; radio stations switched to live television feeds and Web sites stripped out graphics to facilitate the flow of information.

Newspapers across the country printed extra copies and special editions and sold out early in the day on Sept. 12. In addition, the major networks provided continuous coverage and news updates for several days after the attacks. A catastrophe of this magnitude provides an opportunity to understand how people learn about these kinds of events, how rapidly or slowly information spreads and how people respond to the news. This study examines the news diffusion process of the Sept. 11 attacks, as well as individuals' emotional and issue involvement with these events in relationship to media use in the first few days following the attacks. Studies in the news diffusion tradition suggest events that are highly relevant (salient) are diffused throughout the population much more rapidly than events perceived as being of less personal magnitude.

Estimating Terrorism Risk

This research on the general issue of how to estimate terrorism risk is meant to inform resource allocation. It is not a direct assessment of current DHS practice or endorsement of insurance risk modeling. Similarly, it does not represent government policy and is not presented as such. Rather, it is intended to add information and perspective to the rapidly maturing issue of risk-based resource allocation and promote discussion. The Urban Areas Security Initiative (UASI) is a DHS grant program designed to enhance security and overall preparedness by addressing unique planning, equipment, training, and exercise needs of large urban areas. Although there appears to be agreement among many stakeholders that these allocations should reflect the magnitude of risks to which different areas are exposed, no consensus has emerged on how this might be accomplished.

Indeed, the UASI grant program has frequently been criticized for inadequately calculating risk and therefore for failing to distribute resources in proportion to urban areas' shares of total terrorism risk. Ultimately, efficient allocation of homeland security resources would be determined based upon assessment of the cost effectiveness of alternative risk-reduction opportunities. After potentially first addressing obvious and easily mitigated risks, this requires understanding the cost effectiveness of different types and amounts of investment. Neither the methods nor the data are available to answer questions about the effectiveness of available risk-reduction alternatives or to determine reasonable minimum standards for community preparedness.

Until these questions are answered, allocating homeland security resources based on risk is the next best approach since areas at higher risk are likely to have more and larger opportunities for risk reduction than areas at lower risk. That is, resources would be allocated roughly proportionally to the distribution of risk across areas receiving funding. This monograph offers a method for constructing an estimate of city risk shares, designed to perform well across a wide range of threat scenarios and risk types. It also proposes and demonstrates a framework for comparing the performance of alternative risk estimates given uncertainty in measuring the elements of risk.

Components of Risk

Terrorism risk can be viewed as having three components: the *threat* to a target, the target's *vulnerability* to the threat, and the *consequences* should the target be successfully attacked. People and organizations represent threats when they have both the intent and capability to damage a target. The *threats* to a target can be measured as the probability that a specific target is attacked in a specific way during a specified period. Thus, a threat might be measured as the annual probability that a city's football stadium will be subject to attack with a radiological weapon. *Vulnerability* can be measured as the probability that damage occurs, given a threat.

Damages could be fatalities, injuries, property damage, or other consequences; each would have its own vulnerability assessment. *Consequences* are the magnitude and type of damage resulting, given a successful terrorist attack. Risk is a function of all three components: threat, vulnerability, and consequences. These constructs can be used to measure risk consistently in terms of expected annual consequences. More detailed definitions of vulnerability, threat, and risk and discussions of measures for each are presented in this monograph.

Uncertainty and Value Judgments

There are two important sources of uncertainty in estimating terrorism risk. The first includes variability and error in estimates of threats, vulnerabilities, and consequences. The second involves how we should value different types of consequences. Part of an informed discussion of homeland security policy rests on an understandable and transparent means of accounting for uncertainties in estimates and the consequences of using alternative values. When facing uncertainty about estimates and values, policy analysis often relies on best estimates, even when they have a low probability of being correct—and a high probability of being wrong.

While this allows us to generate a very precise estimate of risk, in the end, if the estimates poorly represent what actually happens in real life, the precision is misplaced. So, rather than seek an optimal method for estimating risk, we seek a method that leads us to make the least egregious errors in decision-making across the range of possible scenarios that might develop in the future. Following methods of adaptive planning under deep uncertainty, we seek a method for estimating risk that is robust because it has the lowest expected error when evaluated against a wide range of possible futures.

One approach for developing an estimate with these properties would be first to define multiple sets of threats, vulnerabilities, and consequence measures and use them as the basis for constructing a single risk estimate. Then using these multiple estimates, develop a single description of how risk is distributed that balances across multiple perspectives of terrorism risk. Generating multiple risk estimates can provide plausible bounds on the magnitude of terrorism risk estimates and how different stakeholders may be affected based on where they live or what outcomes they value most. From several estimates, one can ask how low or high terrorism risk may be in a specific city, what a best estimate of risk is given the range of estimates available, and how answers to these questions differ when considering different types of outcomes.

The challenge to analysts defining a single picture of how risk is distributed is to do so without losing significant information. This requires specifying how to deal with the challenges of aggregation given both inherent uncertainties and value choices. Uncertainty in terrorism risk estimates suggests the need to devise means of hedging our homeland security policies against a range of distributions of risk that are plausible given what we know about uncertainties in our risk estimation procedures. So, rather than seek an optimal method for estimating risk, we seek a method that leads us to make the least egregious errors in decision-making across the range of possible scenarios that might develop in the future. This presents a problem comparable to that of forecasting economic trends using multiple estimates or models discussed by Clemen (1989).

This literature highlights two objectives to consider when combining estimates: 1) use information contained in the multiple estimates to improve forecasting accuracy and 2) make note of and retain the important distinctions that individual estimates represent. Addressing multiple values or objectives in terrorism risk estimates differs from combining forecasts. While the goal of combining forecasts is to develop an accurate estimate, the goal of considering multiple objectives is to reflect appropriately the range of values held by stakeholders. Literature on multi-objective decision making provides several approaches for addressing the fact that terrorism risk can be expressed in multiple outcomes. The commonality across these methods is the need to reflect transparently a range of values for multiple objectives in the decision making process.

Despite the many sources of uncertainty surrounding terrorism risk, estimating this risk is necessary for informed distribution of homeland security resources. Approaches that have been used in policy analysis for estimating terrorism risk are bounded by two generic categories: simple risk indicators and event-based models. Each approach reflects the components of terrorism risk and their uncerta ies in different ways. As examples of simple indicators, we describe how population and density-weighted population have been used as estimates of terrorism risk. As an example of event-based models, we describe the Risk Management Solutions (RMS) Terrorism Risk Model. These two examples allow for comparisons that illustrate the strengths and weaknesses of each approach.

There is a logical link between population-based indicators and terrorism risk. An argument can be made that consequences are correlated with population and threats are correlated with population density. There are practical benefits for using simple risk indicators such as those based upon population. In general, the metrics for measuring these indicators are well understood and measurable, and data is widely available. The main limitation of these simple indicators is that they do not fully reflect the interactions of threat, vulnerability, and consequences. As a result, there is little consensus and no validated framework for deciding how to use several simple indicators to create a single risk estimator.

Event-based models are built upon relatively detailed analysis of consequences from specific attack scenarios. These models include sensitivity analysis for important parameters that affect consequences. They may include components to model multiple types of events and multiple targets. They may also include modules that translate expert judgments of likelihood or consequences. The strength of event-based models lie in the greater fidelity they enable in analysis. The weakness is that, to obtain this detail, analysts must estimate many uncertain parameters. One example of an event-based model is the RMS Terrorism Risk Model. The RMS model, discussed in more detail later in this monograph, was developed as a tool for the insurance and reinsurance industries to assess risks of macro terrorism.

Estimating Terrorism Risk

To demonstrate an approach for estimating terrorism risk to inform resource allocations, we calculated a single estimate of U.S. cities' risk shares based on multiple perspectives of terrorism risk obtained from the RMS Terrorism Risk Model. By considering three perspectives on threat (the RMS standard and enhanced and reduced threat outlooks), the RMS results provide three estimates of terrorism risk for the urban areas that received UASI funding in fiscal year 2004. Using these three sets of expected fatalities, we calculated an aggregated risk estimator by minimizing the sum of the squared underestimation error across all urban areas and risk estimates. The assumptions and limitations inherent to this approach are presented in the monograph.

We compared the proposed aggregated risk estimator to simple population-based indicators, looking at both how the distribution of risk changes for each and the propensity for each to underestimate a city's risk share given the uncertainty that surrounds terrorism risk. The aggregated risk estimator concentrated most of the expected terrorism losses in relatively few cities compared to population-based indicators. In addition, the aggregated estimator resulted in the lowest underestimates of risk aggregated across all urban areas.

Density weighted population performed better than population alone and was, in fact, quite a bit closer in performance to that of the aggregated indicator than to that of the simple population indicator. Because the density-weighted population indicator performs well and is easier to derive than the event-based indicator, it might be of utility for some purposes, e.g., in risk-based allocation of resources for strategic purposes over long time intervals, during which relative risk across urban areas is not expected to change much.

Density-weighted population, however, does not allow decision makers to see how changes in threat or vulnerability information affect terrorism risk. For example, when making operational resource allocations or evaluating the effectiveness of preparedness programs, decision makers need to understand how specific countermeasures reduce or change the profile of terrorism risk. Similarly, a crude indicator like density-weighted population would offer no guidance about how city risk estimates might change with, for instance, new intelligence about terrorist targeting or capabilities of using weapons of mass destruction (WMD).

For these purposes, more detailed event based models of terrorism risk are essential. In this study, a single event-based estimate was shown to be robust across uncertainties about the likelihood of WMD attacks, uncertainties about which consequences ought to be prioritized in considerations of city risk, and uncertainties about the expected magnitude of risks each city might face. For example, an important observation is that the risk profile of the urban areas examined did not change significantly with the variability of threats from weapons of mass destruction. This is clearly a function of the models used in this study and how they were parameterized.

While the primary focus of this study was not to estimate precisely terrorism risk in the United States, this observation raises questions such as whether risk is a characteristic of a region's infrastructure or population that is relatively stable across different threats. If so, this would be an important observation when it comes to policy and resource decisions. The Urban Areas Security Initiative (UASI) is a Department of Homeland Security (DHS) grant program designed to enhance security and overall preparedness to prevent, respond to, and recover from acts of terrorism. These goals are accomplished by providing financial assistance to address the unique planning, equipment, training, and exercise needs of large urban areas.

In fiscal year 2004, UASI provided $675 million to 50 urban areas perceived to be at highest risk from terrorist attacks. These funds were allocated using a formula that accounted for several indicators of the terrorism risk to which each urban area might be exposed. Though precise details of the formula are not publicly available, it reportedly calculates each urban area's share of total terrorism risk based on city population, infrastructure, and threat information, giving indicators for each factor an importance weight of nine, six, and three, respectively. Despite this effort to allocate homeland security resources based on the relative risks to which each urban area is exposed, the Department of Homeland Security has frequently been criticized for inadequately calculating risk and therefore for failing to distribute resources in proportion to urban areas' shares of total terrorism risk.

Debates about the proper allocation of resources have suffered from several problems. For instance, currently, there is no shared and precise definition of terrorism risk, so stakeholders in the debate are often referring to different concepts of risk. Even if a precise definition were widely used, there are no standard methods for estimating and monitoring changes in the level and nature of terrorism risks. Instead, various indicators of risk have been used (for instance in the UASI formula) or proposed, which are presumed to correspond in some way with true terrorism risk.

To our knowledge, however, no systematic frameworks for selecting these indicators or aggregating them into a unitary measure of risk are yet available. Moreover, terrorism risk changes over time as terrorist motives, capabilities, and targets change and adapt to risk-mitigation efforts. These facts defy the efficacy of any simplistic model that attempts to enumerate a single index as a measure of risk. Measuring terrorism risk must always reflect uncertainties in estimates of the relative risks faced by different cities.

Risk Assessment versus Resource Allocation

Ultimately, efficient allocation of homeland security resources would be determined based upon assessment of the cost effectiveness of alternative risk-reduction opportunities. This requires understanding the cost effectiveness of different types and amounts of investment. As a hypothetical example, even if terrorism risks were greater in New York City than in Des Moines, Iowa, allocating resources according to proportion of risk would not be optimal if available countermeasures are more cost effective in Des Moines. For example, terrorists could respond strategically to countermeasures in New York City and target less-protected areas, or the marginal effectiveness of resources spent in New York City may decrease with continuing investment.

Neither the methods nor the data are available to answer questions about the effectiveness of available risk-reduction alternatives or to determine reasonable minimum standards for community preparedness. Until these questions are answered, allocating homeland security resources based on risk is the next best approach since areas at higher risk are likely to have more and larger opportunities for risk reduction than are areas at lower risk. That is, resources would be allocated roughly proportionally to the distribution of risk across areas receiving funding. There are several other reasons why it is still important for decision makers to understand the levels and distribution of terrorism risk.

First, because assessing risk and risk reduction is a critical first step in assessing cost effectiveness of counterterrorism efforts, methods developed to support terrorism risk assessment will also support analysis of resource allocation. Further, even when large risks are not mitigated by current efforts, identifying them can help direct intelligence gathering, research, and future counterterrorism efforts. Finally, following changes in the levels and patterns of terrorism risk over time provides insights into the effectiveness of current efforts and the emergence of new risks.

Scope and Limitations

In this monograph, we propose a specific definition of terrorism risk that can be operationalised for practical problems facing DHS and develop a method of constructing a single measure of risk that accounts for uncertainties in risk measurement. We then propose and demonstrate a framework for evaluating this measure, along with alternative measures of risk, to understand resulting errors given uncertainties in their measurement. Finally, we offer recommendations on future efforts to calculate the shares of total terrorism risk to which different areas are exposed. While the discussions in this monograph focus on a specific program, the UASI grants, the problems discussed previously are common to a number of risk estimation problems in homeland security. Thus, the problem is a general one of decision-making under uncertainty, and the solutions presented here are also generalizable to similar decision contexts.

This monograph does not address all problems identified previously. This treatment of risk estimation does not fully inform specification of a formula for risk-based allocation of homeland security resources. As noted previously, before such a formula can be constructed, additional research is needed to understand the relationship between resource allocation and risk reduction. The scope of this project is further limited to the direct consequences associated with terrorism threats. Thus, we do not include in our estimates of terrorism risk the secondary and higher order economic or other losses that result from an attack on a given location.

These effects are important and may well constitute the major portion of the risk but can be handled by the methods we develop here given additional resources. Such analysis would extend this current work to further improve the evaluation of the relative risks to which different urban areas are exposed and thus would help to improve the allocation of homeland security resources. Differing notions of terrorism risk frequently fuel disagreements about the relative risks to which different regions or cities are exposed. Some arguments implicitly link risk to terrorism threats. If, for example, one city were known through gathered intelligence or past history to be the preferred target for terrorists, this view would support a claim that this city has a high level of terrorism risk.

Alternatively, others argue that risk is more closely associated with infrastructure vulnerabilities within a region because these represent logical targets for terrorism. Thus, for example, even if we do not know of a threat to a nuclear power plant, reason and prudence argue that we should include that facility in considering a region's risk. Finally, discussions of risk occasionally emphasize the possible consequences of terrorist attacks in evaluating risk. Thus, if two cities have similar chemical storage facilities, but one has the facility located close to its population center, a persuasive argument can be made that the nearer-to-population city's chemical facility presents a greater risk than the other cities.

Clearly, strong arguments can be made that threats, vulnerabilities, and consequences play a significant part in the overall risk to which a city is exposed. What has been less clear is how these three components are related. In this chapter, we offer a definition of risk that links them. We also distinguish between definitions for threat, vulnerability, consequences, and risks and the measures that can be used to assess and track each.

Threat

People or organizations represent a terrorist threat when they have the intent and capability to impose damage to a target. Note that neither intentions without capabilities nor capabilities without intentions pose a threat. Threat only exists when both are manifested together in a person or organization. Allocating homeland security resources to protect critical infrastructure or cities requires measuring the threats posed to specific targets or from specific types of attack. When the scope of threat is defined in terms of a specific set of targets, a specific set of attack types, and a specific time period, probability can be used as a measure of the likelihood that an attack will occur.

Measure Threat

This measure of terrorist threat emphasizes a specific type of attack on specific targets. Radiological attack represents a different threat to a specific target than nuclear attack. Attacks on stadiums represent different threats than attacks on skyscrapers. A complete description of the threats to which a target is exposed would require consideration of every mode of attack separately. In practice, however, it may suffice to focus on a limited number of attack types that are representative of chemical, biological, radiological, or nuclear (CBRN) and explosive attack modes. Similarly, it may suffice to focus on a limited number of target types or groups of targets in a region.

This measure of threat is specified in terms of attack types and targets. The intelligence community more customarily considers threat in terms of groups of attackers given its interest identifying and stopping those who might pose a threat. An attack-type perspective is more useful for the task of resource allocation because the decision context is most concerned with what targets are threatened than with by whom and why. Finally, since our measure for threat is uncertain, one should keep in mind that it can also be represented by a probability distribution, not a point estimate. These definitions are consistent with methods and terminology proposed through applications of engineering risk analysis to terrorism risk assessment.

Vulnerability

Clearly, not all threats of the same type are equally important. Furthermore, the threat of terrorism is dynamic in that it adapts to current conditions that affect the likelihood of attack success. For example, even if a typical hotel and fortified military base have equal probability of being subjected to a car-bomb attack, the attack would be more likely to achieve the aim of causing significant damage at the less-secure hotel. Therefore, we also need a precise definition of vulnerability that captures information about the infrastructure in which we are interested. Paraphrasing Haimes, *vulnerability is the manifestation of the inherent states of the system (e.g., physical, technical, organizational, cultural) that can result in damage if attacked by an adversary.*

Referring again to the domain of engineering risk analysis, where threat can be thought of as being a load or force acting on a system, vulnerability can be thought of as being the capacity of a system to respond to this threat. To use this definition for measurement, we must be more specific and ask, "vulnerable to what?" Probability can be used as a measure of the likelihood that vulnerability will lead to damage when attacks occur. In other words, a target's vulnerability can be articulated as the probability that an attack of a given type will be successful once it has been launched and, as articulated, measures vulnerability to specific types of damages only (i.e., there would be separate vulnerability assessments for deaths, injuries, and property damage).

Note that for the measure specified above, magnitude of the damage is not part of the definition of vulnerability. This measure assumes a simplified representation of vulnerability in which there is either a successful attack with damage or no success with no damage. As a result, we define "success" in terms of whether or not damage, having a distribution of magnitude, is inflicted by the attack. Consequence measurement is discussed below. A more general model (used in many military analyses) is that there is a range of damage levels, each associated with its own probability. This is simply a more discrete representation of damage and defense mechanisms.

Consequences

We define "consequence" as the magnitude and type of damage resulting from successful terrorist attacks. To define a measure of consequence, specificity is again required. In this case, specificity requires treatment of two important considerations: how consequences are measured and how uncertainty is addressed.

Measure Consequence

Consequences can be expressed in terms of fatalities, injuries, economic losses, or other types of damage. Other aspects of consequences can also be considered using the approach we outline here and this definition. For example, the damage or destruction of critical infrastructures that cause injury, loss of life, and economic damage outside the area of immediate attack are important. They may in fact dominate the results of an analysis if the impact on society as a whole is considered rather than the impact on the target and its occupants and owners.

In this monograph, however, we limit our focus to mortality, morbidity, and economic loss at the point of attack in order to illustrate an approach to risk estimation in a manner that is transparent yet relevant to real-world policy decisions. Consequences are determined by many uncertain factors, such as wind speed or relative humidity (which could be important factors in a chemical or biological attack, for example). These uncertainties can be addressed by considering a full distribution for potential consequences or specific points along this distribution.

Haimes notes that risk assessment of rare and extreme events requires special consideration of the tails of these distributions, and that the expected value often misrepresents true risk. Conversely, estimates of the tail of the distribution will be very dependent upon assumptions when considering events like terrorism where there is great uncertainty about events and limited historical information. For this reason, and to simplify, our continued discussion of consequences considers the expected value of the distribution of damage.

Vulnerability and Consequences

Risk is the anticipated consequences over some period of time to a defined set of targets, resulting from a defined set of threats. In other words, terrorism risk represents the expected consequences of attacks taking into account the likelihood that attacks occur and that they are successful if attempted. In probabilistic terms, risk from an attack of a certain type is the unconditional expected value of damages of a certain type. There are two advantages of using this formulation of terrorism risk. First, it provides an approach for comparing and aggregating terrorism risk. With this definition, it is possible to compare risks of a specific type across diverse targets such as airports and electrical substations. For example, the injury risk from an explosives attack could be expressed for both as the expected annual injuries resulting from such attacks against each target.

Estimating overall terrorism risk requires further analysis that considers all threat types and targets. If risks were independent, expected damages of a specific type could be aggregated by summing across threat types and target types. However, dependencies likely exist between risks. For instance, a successful nuclear attack in a city could dramatically change the expected risks for targets in the damage footprint of the explosion. Second, this definition of risk provides a clear mapping between risk and approaches to managing or reducing risk. Intelligence and active defense involving "taking the fight to the enemy" represent an approach to risk management that focuses specifically on threats.

Managing risk through vulnerability requires increasing surveillance and detection, hardening targets, or other capabilities that might reduce the success of attempted attacks. Finally, risk can be managed through consequences by increasing preparedness and response that reduces the effects of damage through mitigation or compensation.

6

Media Coverage Fuels
Terrorism

Terrorists, governments, and the media see the function, roles and responsibilities of the media when covering terrorist events from differing and often competing perspectives. Such perspectives drive behavior during terrorist incidents—often resulting in both tactical and strategic gains to the terrorist operation and the overall terrorist cause. The challenge to both the governmental and press communities is to understand the dynamics of terrorist enterprise and to develop policy options designed to serve the interests of government, the media, and the society. More ink equals more blood, claim two economists who say that newspaper coverage of terrorist incidents leads directly to more attacks. It's a macabre example of win-win in what economists call a "common-interest game," say Bruno S. Frey of the University of Zurich and Dominic Rohner of Cambridge University.

"Both the media and terrorists benefit from terrorist incidents," their study contends. Terrorists get free publicity for themselves and their cause. The media, meanwhile, make money "as reports of terror attacks increase newspaper sales and the number of television viewers." The researchers counted direct references to terrorism between 1998 and 2005 in the New York Times and Neue Zuercher

Zeitung, a respected Swiss newspaper. They also collected data on terrorist attacks around the world during that period. Using a statistical procedure called the Granger Causality Test, they attempted to determine whether more coverage directly led to more attacks.

The results, they said, were unequivocal: Coverage caused more attacks, and attacks caused more coverage — a mutually beneficial spiral of death that they say has increased because of a heightened interest in terrorism since Sept. 11, 2001. One partial solution: Deny groups publicity by not publicly naming the attackers, Frey said. But won't they become known anyway through informal channels such as the Internet? Not necessarily, Frey said. "Many experiences show us that in virtually all cases several groups claimed responsibility for a particular terrorist act. I would like the same rule that obtains within a country: Nobody can be called a criminal — in our case a terrorist — if this has not been established by a court of law."

While I've never heard of Rohner, I'm quite familiar with Frey's work in political economy; he's highly regarded. The Granger Causality Test is well outside the scope of my methodological expertise but its creator won a Nobel Prize, which leads me to think there's something to it. This is a case where sophisticated research produces results that match up with our intuition. It's no secret that media coverage is a prime motivation of terrorists, if not the primary motivation at the tactical level. It's hard to sow terror if the results of one's carnage are only known by eyewitnesses. Nor is it surprising that, as terrorist strikes increase, coverage goes up.

Unfortunately, Frey's solution is a non-starter. Even aside from 1st Amendment concerns, a ban on calling terrorists "terrorists"–which many media outlets have already self-imposed–would do nothing to forestall coverage of terrorist activity. It's the carnage and ensuing terror, not the label "terrorist," that the perpetrators are after.

Terrorists must have publicity in some form if they are to gain attention, inspire fear and respect, and secure favorable understanding of their cause, if not their act. Governments need public understanding, cooperation, restraint, and loyalty in efforts to limit terrorist harm to society and in efforts to punish or apprehend those responsible for terrorist acts. Journalists and the media in general pursue the freedom to cover events and issues without restraint, especially governmental restraint.

Three new trends appear to be emerging which impact on the relationship between the media, the terrorist, and government. These include: (1) anonymous terrorism; (2) more violent terrorist incidents; and (3) terrorist attacks on media personnel and institutions.

A number of options, none without costs and risks, exist for enhancing the effectiveness of government media-oriented responses to terrorism and for preventing the media from furthering terrorist goals as a byproduct of vigorous and free reporting. These include: (1) financing joint media/government training exercises; (2) establishing a government terrorism information response center; (3) promoting use of media pools; (4) promoting voluntary press coverage guidelines; and (5) monitoring terrorism against the media.

The media and the government have common interests in seeing that the media are not manipulated into promoting the cause of terrorism or its methods. But policymakers do not want to see terrorism, or anti-terrorism, eroding freedom of the press—one of the pillars of democratic societies. This appears to be a dilemma that cannot be completely reconciled—one with which societies will continually have to struggle. The challenge for policymakers is to explore mechanisms enhancing media/government cooperation to accommodate the citizen and media need for honest coverage while limiting the gains uninhibited coverage may provide terrorists or their cause. Communication between the government and the media here is an important element in any strategy to prevent terrorist causes and strategies from prevailing and to preserve democracy.

This responds to a range of inquiries received by CRS on the nature of the relationship of terrorist initiatives, publicity, and governments. The media are known to be powerful forces in confrontations between terrorists and governments. Media influence on public opinion may impact not only the actions of governments but also on those of groups engaged in terrorist acts. From the terrorist perspective, media coverage is an important measure of the success of a terrorist act or campaign. And in hostage-type incidents, where the media may provide the only independent means a terrorist has of knowing the chain of events set in motion, coverage can complicate rescue efforts.

Governments can use the media in an effort to arouse world opinion against the country or group using terrorist tactics. Public diplomacy and the media can also be used to mobilize public opinion in other countries to pressure governments to take, or reject, action against terrorism. Margaret Thatcher's metaphor that publicity is the oxygen of terrorism underlines the point that public perception is a major terrorist target and the media are central in shaping and moving it. For terrorism, the role of the media is critical.

This report examines competing perspectives on the desired role for the media when covering terrorist incidents: what the terrorist wants, what the government wants, and what the media wants when covering a terrorist event. These are classic perspectives drawn from the experiences of this century. It then addresses three recent trends that impact on the relationship between terrorism and the media and concludes with options for congressional consideration.

Terrorists, governments, and the media see the function, roles and responsibilities of the media, when covering terrorist events, from differing and often opposing perspectives. Such perceptions drive respective behaviors during terrorist incidents—often resulting in tactical and strategic gains, or losses, to the terrorist operation and the overall terrorist cause. The challenge to the governmental and press community is to understand the dynamics of terrorist enterprise and to develop policy options to serve government, media and societal interests.

What Terrorists Want from Media

- Terrorists need publicity, usually free publicity that a group could normally not afford or buy. Any publicity surrounding a terrorist act alerts the world that a problem exists that cannot be ignored and must be addressed. From a terrorist perspective, an unedited interview with a major figure is a treasured prize, such as the May 1997 CNN interview with Saudi dissident, terrorist recruiter and financier Usama Bin Ladin. For news networks, access to a terrorist is a hot story and is usually treated as such.

- They seek a favourable understanding of their cause, if not their act. One may not agree with their act but this does not preclude being sympathetic to their plight and their cause. Terrorists believe the public "needs help" in understanding that their cause is just and terrorist violence is the only course of action available to them against the superior evil forces of state and establishment. Good relationships with the press are important here and they are often cultivated and nurtured over a period of years.

- Terrorist organizations may also seek to court, or place, sympathetic personnel in press positions—particularly in wire services—and in some instances may even seek to control smaller news organizations through funding.

- Legitimacy - Terrorist causes want the press to give legitimacy to what is often portrayed as ideological or personality feuds or divisions between armed groups and political wings. For the military tactician, war is the continuation of politics by other means; for the sophisticated terrorist, politics is the continuation of terror by other means. IRA and Hamas are examples of groups having "political" and "military" components. Musa Abu Marzuq, for example, who was in charge of the political wing of Hamas is believed to have approved specific bombings and assassinations. Likewise, the "dual hat" relationship of Gerry Adams of Sinn Fein—the purported political wing of the IRA—to other IRA activities is subject to speculation. Distinctions are often designed to help people join the ranks, or financially contribute to the terrorist organization.

- They also want the press to notice and give legitimacy to the findings and viewpoints of specially created non-governmental organizations (NGOs) and study centers that may serve as covers for terrorist fund raising, recruitment, and travel by terrorists into the target country. The Palestinian Islamic Jihad-funded and controlled World and Islam Studies Enterprise is but one known example. The Hamas-funded Islamic Association for Palestine (LAP) in Richardson, Texas, is another of many.

- In hostage situations—terrorists need to have details on identity, number and value of hostages, as well as details about pending rescue attempts, and details on the public exposure of their operation. Particularly where state sponsors are involved, they want details about any plans for military retaliation.

- Terrorist organizations seek media coverage that causes damage to their enemy. This is particularly noticeable when the perpetrators of the act and the rationale for their act remain unclear. They want the media to amplify panic, to spread fear, to facilitate economic loss (like scaring away investment and tourism), to make populations loose faith in their governments' ability to protect them, and to trigger government and popular overreaction to specific incidents and the overall threat of terrorism.

What Government Leaders Want from the Media

Governments seek understanding, cooperation, restraint, and loyalty from the media in efforts to limit terrorist harm to society and in efforts to punish or apprehend those responsible for terrorist acts, specifically:

- They want coverage to advance their agenda and not that of the terrorist. From their perspective, the media should support government courses of action when operations are under way and disseminate government provided information when requested. This includes understanding of policy objectives, or at least a balanced presentation, e.g., why governments may seek to mediate, yet not give in to terrorist demands.

- An important goal is to separate the terrorist from the media—to deny the terrorist a platform unless to do so is likely to contribute to his imminent defeat.

- Another goal is to have the media present terrorists as criminals and avoid glamorizing them; to foster the viewpoint that kidnapping a prominent person, blowing up a building, or hijacking an airplane is a criminal act regardless of the terrorists' cause.

- In hostage situations, governments often prefer to exclude the media and others from the immediate area, but they want the news organizations to provide information to authorities when reporters have access to the hostage site.

- They seek publicity to help diffuse the tension of a situation, not contribute to it. Keeping the public reasonably calm is an important policy objective.

- It is generally advantageous if the media, especially television, avoids "weeping mother" emotional stories on relatives of victims, as such coverage builds public pressure on governments to make concessions.

- During incidents, they wish to control terrorist access to outside data—to restrict information on hostages that may result in their selection for harm; government strongly desires the media not to reveal planned or current anti-terrorist actions or provide the terrorists with data that helps them.

- After incidents, they want the media not to reveal government secrets or detail techniques on how successful operations were performed—and not to publicize successful or thwarted terrorist technological achievements and operational methods so that copycat terrorists do not emulate or adapt them.

- They want the media to be careful about disinformation from terrorist allies, sympathizers, or others who gain from its broadcast and publication. Many groups have many motives for disseminating inaccurate or false data, including, for example, speculation as to how a plane may have been blown up, or who may be responsible.

- They want the media to boost the image of government agencies. Agencies may carefully control leaks to the press giving scoops to newsmen who depict the agency favorably and avoid criticism of its actions.

- They would like journalists to inform them when presented with well grounded reasons to believe a terrorist act may be in the making or that particular individuals may be involved in terrorist activity.

- In extreme cases, where circumstances permit, vital national security interests may be at stake, and chances of success high, they may seek cooperation of the media in disseminating a ruse that would contribute to neutralizing the immediate threat posed by terrorists. In common criminal investigations involving heinous crimes, such media cooperation is not uncommon—when media members may hold back on publication of evidence found at a crime scene or assist law enforcement officials by publishing misleading information or a non-promising lead to assist authorities in apprehending a suspect by, for example, lulling him or her into a false sense of security.

The Media Covering Terrorist Incidents or Issues

Journalists generally want the freedom to cover an issue without external restraint—whether it comes media owners, advertisers, editors, or from the government.

- Media want to be the first with the story. The scoop is golden, "old news is no news." Pressure to transmit real time news instantly in today's competitive hi-tech communication environment is at an all-time high.

- The media want to make the story as timely and dramatic as possible, often with interviews, if possible. During the June 1985 TWA Flight 847 hijack crisis, ABC aired extensive interviews with both hijackers and hostages. (A photo was even staged of a pistol aimed at the pilot's head.)

- Most media members want to be professional and accurate and not to give credence to disinformation, however newsworthy it may seem. This may not be easily done at times, especially when systematic efforts to mislead them are undertaken by interested parties.

- They want to protect their ability to operate as securely and freely as possible in the society. In many instances, this concern goes beyond protecting their legal right to publish relatively unrestrained; it includes personal physical security. They want protection from threat, harassment, or violent assault during operations, and protection from subsequent murder by terrorists in retaliation providing unfavorable coverage (the latter occurring more often abroad than in the United States.)

- They want to protect society's right to know, and construe this liberally to include popular and dramatic coverage, e.g., airing emotional reactions of victims, family members, witnesses, and "people on the street," as well as information withheld by law enforcement, security, and other organs of government.

- Media members often have no objection to playing a constructive role in solving specific terrorist situations if this can be done without excessive cost in terms of story loss or compromise of values.

New Trends Impacting

A series of recent terrorist acts indicates the emergence of trends that impact on the relationship between the media, the terrorist, and government. These include: (1) a trend toward anonymity in terrorism; (2) a trend towards more violent terrorist incidents; and (3) a trend towards attacks on media personnel and institutions.

Today we see instances of anonymous terrorism where no one claims responsibility and no demands are made. The World Trade Center bombing is but one example. This allows the media a larger role in speculation, and generally removes most basis for charges that they are amplifying a terrorist's demands or agenda. Reportage is inevitable; especially if it includes unbridled speculation, false threats or hoaxes, coverage can advance terrorists' agendas, such as spreading panic, hurting tourism, and provoking strong government reactions leading to unpopular measures, including restrictions on individual liberties.

In the context of advanced information and technology, a trend suggesting more violent terrorism cannot be ignored. The Department of State's *Patterns of Global Terrorism: 1996* notes that while worldwide instances of terrorist acts have dropped sharply in the last decade, the death toll from acts is rising and the trend continues "toward more ruthless attacks on mass civilian targets and the use of more powerful bombs. The threat of terrorist use of materials of mass destruction is an issue of growing concern...". If, and as, terrorism becomes more violent, perceptions that the press is to some degree responsible for facilitating terrorism or amplifying its effects could well grow. Increasingly threatened societies may be prone to take fewer risks in light of mass casualty consequences and may trust the media less and less to police itself.

Attacks on Media Personnel

Attacks on journalists who are outspoken on issues of concern to the terrorists seem to be on the rise. Recent attacks occurred in Algeria, Mexico, Russia, Chechnya, and London, but there have been cases as well in Washington, D.C. at the National Press Building and at the United Nations in New York. One private watchdog group estimates that forty-five journalists were killed in 1995 as a consequence of their work.

A number of options might be considered to improve government/media interaction when responding to or covering terrorist incidents. These include: (1) financing joint media/government training exercises; (2) establishing a government terrorism information response center; (3) promoting use of media pools for hostage-centered terrorist events; (4) establishing and promoting voluntary press coverage guidelines; and (5) monitoring terrorism against the media.

Effective public relations usually precedes a story— rather than reacts to it. Nations can beneficially employ broad public affairs strategies to combat terrorist-driven initiatives, and the media can play an important role within the framework of such a strategy. Training exercises are vital: exercises such as those conducted by George Washington University and the Technology Institute in Holon, Israel, which bring together government officials and media representatives to simulate government response and media coverage of mock terrorist incidents. Promoting and funding of similar programs on a broad scale internationally is an option for consideration.

One option Congress might consider would be establishment of a standing government terrorist information response center (TIRC). Such a center, by agreement with the media, could have on call (through communication links) a rapid reaction terrorism reporting pool composed of senior network, wire-service, and print media representatives. Network coverage of incidents would then be coordinated by the network representative in the center. Such a center could be headed by a government spokesperson (the Terrorism Information Coordinator, TIC) who could seek to promptly seize the information and contexting initiative from the particular terrorist group.

Too often, when incidents happen in the United States there is a vacuum of news other than the incident itself, and by the time the government agencies agree on and fine tune what can be said and what positions are to be taken, the government information initiative is lost.

Another option that has been mentioned specifically for coverage of hostage type events, would be use of a media pool where all agree on the news for release at the same time. A model would need to be established. However, media agreement would not be easily secured.

Another option would be establishment by the media of a loose code of voluntary behavior or guidelines that editors and reporters could access for guidance. Congress could urge the President to call a special media summit, national or perhaps international in scope under the anti-terrorism committed G-8 industrialized nations summit rubric, for senior network and print media executives to develop voluntary guidelines on terrorism reporting. Another option might be to conduct such a national meeting under the auspices of a new government agency.

Areas for discussion might be drawn from the practices of some important media members and include guidelines on:

- Limiting information on hostages which could harm them: e.g., number, nationality, official positions, how wealthy they may be, or important relatives they have;
- Limiting information on military, or police, movements during rescue operations;
- Limiting or agreeing not to air live unedited interviews with terrorists;
- Checking sources of information carefully when the pressure is high to report information that may not be accurate—as well as limiting unfounded speculation;
- Toning down information that may cause widespread panic or amplify events which aid the terrorist by stirring emotions sufficiently to exert irrational pressure on decision makers.

Even if specific guidelines were not adopted, such a summit would increase understanding in the public policy and press policy communities of the needs of their respective institutions.

Tracking Terrorism against the Media

Finally, a trend toward terrorist attacks against media personnel and institutions may be emerging. This issue was addressed by President Clinton in a meeting with members of the press in Argentina during a state visit there October 17, 1997, when the President expressed concern over the issue of violence and harassment of the press in Argentina and suggested that the Organization of American States (OAS) create a special unit to ensure press freedom similar to the press ombudsman created by the Organization on Security and Cooperation in Europe (OSCE). Notwithstanding, comprehensive and readily available government statistics are lacking. One way to approach this problem would be for government reports on terrorism, such as the U.S. Department of State's *Patterns of Global Terrorism*, to include annual statistics showing the number of journalists killed or injured yearly in terrorist attacks and the annual number of terrorist incidents against media personnel or media institutions.

The media and the government have common interests in seeing that the media are not manipulated into promoting the cause of terrorism or its methods On the other hand, neither the media or policymakers want to see terrorism, or counter terrorism, eroding constitutional freedoms including that of the press—one of the pillars of democratic societies. This appears to be a dilemma that cannot be completely reconciled—one with which U.S. society will continually have to struggle.

Communication between the government and the media is an important element in any strategy designed to prevent the cause of terrorism from prevailing and in preserving democracy. By their nature, democracies with substantial individual freedoms and limitations on police powers offer terrorists operational advantages. But terrorists and such democracies are not stable elements in combination. If terrorism sustains itself or flourishes, freedoms shrink, and in societies run by ideological authoritarians, thugs, or radical religious extremists, a free press is one of the first institutions to go.

Uncertainty and Values in Terrorism Risk Assessment

The reality that threat, vulnerability, and consequences are all subject to tremendous uncertainties makes estimating each a challenging task. To facilitate risk estimation, it is important to understand the sources of these uncertainties that affect terrorism risk. There are two important sources of uncertainty in estimating terrorism risk. The first includes variability and error in estimates of threats, vulnerabilities, and consequences. For example, exact knowledge of the threat would require comprehensive intelligence on the plans and capabilities of all terrorist groups. Since this level of precision is not feasible, expert judgments must be substituted for fact, resulting in parameter estimates for threats that are subject to uncertainty or frank dis-agreements.

The second source of uncertainty concerns how we should value different types of consequences. This is a fundamental problem underlying homeland security decisions that inevitably share burdens of cost and risk among different parts of U.S. society. For instance, if a city has a small property-loss risk, but a large fatality risk, you might conclude that it has medium overall terrorism risk by valuing the importance of each type of consequence as roughly equivalent. In fact, any strategy for judging the relative importance of different types of consequence represents an attempt to estimate the value that we should or does place on each consequence. Because this requires value judgments—and potentially contentious ones—it must ultimately be discussed by the public and policymakers. Part of an informed discussion of this judgment rests on an understandable and transparent illustration of the consequences of using alternative values.

At any moment, we may assume that a target has exact, true values for threat, vulnerability, consequences and therefore for risk. These values, however, cannot be directly observed, so must be estimated. Estimation introduces uncertainty and error. Probably the greatest source of uncertainty derives from estimates of threat, which concerns terrorists' goals, motives, and capabilities. Our sources of information on these factors—chiefly intelligence, historical analysis, and expert judgment—support only crude estimates of the probability of attacks against specific targets or classes of targets (e.g., banks). Experts frequently disagree about the goals of terrorist groups and their capabilities, and some terror groups may exist about which little is known.

Consequently, assessments of terrorist motivations and capabilities may systematically under- or overestimate threats. Given this, our threat estimates must be treated with suspicion. Vulnerability estimates may be subject to lower levels of uncertainty. Because vulnerability concerns the likelihood that an attack of a specific type and magnitude will be successful against a target, it concerns matters that can, in principle, be carefully studied and for which rough estimates may be reasonably good, e.g., the methods of engineering risk analysis that have been used successfully in estimating risks of space flight and operating nuclear reactors. These methods can be applied to protecting targets and infrastructure (Haimes). Despite the applicability of such approaches, imprecision remains in the estimation of a target's vulnerability.

The sources of uncertainty in consequences concern damage assessments that depend on the physical situation at the target when the attack occurs. For example, suppose a chemical weapon that takes the form of a spray of fine droplets is used at a popular oceanfront recreational location on a busy weekend afternoon and that the chemicals are only effective while they remain in droplet form—that is, they are not effective after they evaporate. If the atmospheric conditions are hot and dry, and the wind is blowing from the land out over the sea, then the droplets might evaporate more quickly due to the dry heat and will blow out over the water away from areas where many people gather.

However, if the weather conditions are such that the wind is blowing over the beach on a humid day, the damage from fatalities and injuries might be more severe. This example illustrates that estimating consequences requires a substantial amount of work. Fortunately, this work is often in the form of straightforward engineering and statistical problems, and well-developed models exist of many natural disasters that are directly applicable, or nearly so. Additionally, the military and other government agencies have long studied the effects of weapons on people and structures, and this, too, is useful for estimating consequences. For example, modeling has been used to estimate the consequences of attacks such as releases of hazardous chemicals or biological agents, radiological bombs, conventional explosions, and nuclear detonations.

Other approaches, like the Interoperability Input-Output model, use economic data to understand the indirect economic consequences of attacks resulting from interdependence between market sectors. Although more information is available on many consequences, the precision with which that information may be applied should not be overstated.

Reflecting Values in Terrorism Risk Management

Models of terrorist attacks can assess impacts in terms of injuries and fatalities, property loss, economic losses, citizen confidence and feelings of security, or myriad other potentially relevant outcomes. Risk can likewise be expressed in terms of any one, or a combination, of these consequences. The emphasis placed on each type of consequence in the evaluation of terrorism risk is a value judgment. While all types of risk could mathematically be combined into a single-dimensional aggregate risk, any such aggregation requires making value judgments on the relative importance of different consequences. This multidimensional nature of consequences creates difficult decisions for policymakers who must weigh the relative importance of different types of consequences when allocating homeland security resources.

As an example, consider two regions. The first, a densely populated business district, may be viewed as a terrorist target for being a hub of economic activity. Terrorist attacks at this location could produce large numbers of fatalities and vast economic losses. The second region, an industrial park where petroleum refining and transferring take place, could also be considered a target. For this case, economic losses may be equally large; however, because of lower population densities, expected fatalities might be lower.

The priorities given to these two regions in estimating risk are driven by values—in other words, the relative weight assigned to a particular type of consequences. Different stakeholder groups will have different perspectives. Some may believe that mitigating risk is exclusively about the prevention of deaths and injuries, and thus the business district bears the greatest burden of risk. Another group may value minimizing damage to the domestic economy more, and consequently believe that risk is divided more equally between the two regions. In our system of allocating homeland security resources, decision makers have an important role in understanding and representing diverse sets of values. In the end, the approach used in the decision-making process should allow transparency, so that citizens can effectively participate in risk-management deliberations.

The uncertainties described earlier ensure there will be disagreement on how to model terrorist events and protection strategies, how to specify probability distributions to represent threat or vulnerability, and how to value diverse measures of consequences. Lempert, Popper, and Bankes (2003) define these conditions as reflecting a state of deep uncertainty. Deep uncertainty has important implications for decision makers charged with developing policies that depend on terrorism risk assessments, like risk-based allocation formulae. Often policy analysis relies on best estimates even when they have a low probability of being correct—and a high probability of being wrong.

While this generates a very precise estimate of risk, in the end, if the estimates poorly represent what actually happens in real life, the precision is misplaced. Lempert, Popper, and Bankes (2003) propose an alterative approach when addressing conditions of deep uncertainty. The approach specifies a wide range of future scenarios that could unfold and then challenges decision makers to choose strategies that perform well across a large number of these possible futures, rather than for a single best estimate of the future.

This approach is consistent with methods of capabilities-based planning that the Department of Defense has adopted to ensure adaptive ness in response to uncertainty about future threats. By analogy, if we believed some model of the stock market was very reliable or subject to low rates of error, then we would allocate most or all of our investment in those stocks it predicts to rise. That is, we would base our policy chiefly on the model's best estimates. In contrast, if we believed the same estimates were subject to considerable error, we would be wise to hedge our investments against alternative possible market outcomes within the range of plausible futures suggested by the uncertainty around our model estimates.

Similarly, in this chapter, we argue that uncertainty in terrorism risk estimates suggests the need to devise means of hedging our homeland security policies against a range of distributions of risk that are plausible given what we know about uncertainties in our risk-estimation procedures. So, rather than seek an optimal method for estimating risk, we seek a method that leads us to make the least egregious errors in decision-making across the range of possible scenarios that might develop in the future. In other words, following Lempert, Popper, and Bankes (2003), when confronting deep uncertainty we seek a method for estimating risk that is robust because it has the lowest expected error when evaluated against a wide range of possible futures.

This decision rule is conceptually similar to Savage's minimax principle and regret minimization. One approach for developing an estimate with these properties would be first to define multiple sets of threats, vulnerabilities, and consequence measures and use them as the basis for constructing a single risk estimate. Then, using these multiple estimates, one could develop a single description of how risk is distributed that balances across multiple perspectives of terrorism risk. This estimate would describe the distribution of aggregate risk and would potentially be robust across a wide range of possible futures and values. That is, an estimate may be designed to minimize the error between estimated risk and actual risks that may materialize as the future unfolds.

Each estimate of threats, vulnerabilities, and consequences represents a different view of what is valued and likely to take place in the future. These views could be results from multiple parameterizations of a single model, results from multiple models, or the perspectives of different experts. The different views could produce estimates of the concentration of terrorism risk in different geographic areas, through different modes of attack, or for different types of consequences. For example, in each view, a different region might be found to have the greatest share of terrorism risk, or consequences may be measured using a different outcome. Generating multiple risk estimates provides information about the bounds of understanding about terrorism risk and how different stakeholders may be affected based on where they live or what out comes they value most.

From several estimates, one can ask how low or high terrorism risk may be in a specific city, what a best estimate of risk is given the range of estimates available, and how answers to these questions differ when considering different types of outcomes. The challenge to analysts defining a single picture of how risk is distributed is to do so without losing significant information. This requires specifying how to deal with the challenges of aggregation given both inherent uncertainties and value choices discussed previously.

Aggregating Risk Estimates

The previous discussion suggests that uncertainty in terrorism risk estimates can be addressed through simple aggregations of multiple perspectives on threat. In fact, this approach has proved useful in another field, economic forecasting. Combining economic forecasts is conceptually similar to combining estimates of terrorism risk. In each case, there is uncertainty about future events that increases as one focuses further into the future. For both terrorism and economic forecasting, competing models exist built on different assumptions about model structure and parameters. In several cases, the competing models for each are highly correlated. Clemen's (1989) literature review demonstrates that combining economic forecasts can improve the accuracy of predictions and those simple aggregations of estimates, such as averaging, can perform well when compared to more complex methods that take into account correlations between estimates or judgments of forecast quality.

Of course, the hazard of aggregating forecasts is that important divergent perspectives might be lost. For example, Morgan and Keith (1995) demonstrated that expert judgments about climate change reveal tremendous diversity of opinion. Thus, it is important to achieve two goals when combining forecasts: 1) achieving the potential gains in accuracy that come from multiple forecasts and 2) make note of and retain the important distinctions that individual forecasts represent.

Values and Multi-objective Decision-making

Addressing multiple values or objectives in terrorism risk estimates differs from combining forecasts. While the goal of combining forecasts is to develop an accurate estimate, the goal of considering multiple objectives is to reflect appropriately the range of values held by stake-holders. For example, a terrorism risk estimation model may yield estimated losses in terms of fatalities and economic damage. Using the estimate of fatalities is not more accurate or inaccurate than using the estimate of economic damages. Rather, it is important that decision makers understand and consider the tradeoffs across consequence types (or objectives) when selecting among policy alternatives.

Literature on multi-objective decision-making provides several approaches for addressing the fact that terrorism risk can be expressed in multiple outcomes. Methods such as multi attribute utility analysis and multi objective value models emphasize the need to structure decisions, elicit stakeholder preferences, and apply axiomatic rules for combining outcomes. Tradeoff analysis reduces the burden of the elicitation process by identifying dominating alternatives and eliminating inferior choices through specification of equivalent choices across objectives. Hierarchical Holographic Modeling incorporates multiple objectives by capturing and representing alternative views of a given problem for decision makers. The commonality across these methods is the need to reflect transparently a range of values for multiple objectives in the decision making process.

Though multi-objective decision making is an essential part of the policy process, the remainder of this monograph provides a demonstration of different approaches to estimating risk and combining estimates without attempting to characterize stakeholders' values or preferences. Instead, we only discuss risk in terms of expected fatalities. Thus, to the extent that risks are distributed differently if measured as injuries or economic consequences, this becomes an additional source of uncertainty in our estimation of overall risk. A comprehensive terrorism risk assessment must allow decision makers to understand the implications of different value judgments. Thus, estimates presented in subsequent chapters of this monograph must be interpreted with this limitation in mind.

Terrorism Risk in Urban Areas

Despite the many sources of uncertainty surrounding terrorism risk, estimating this risk is necessary for informed distribution of homeland security resources. This chapter describes two approaches for estimating terrorism risk: simple risk indicators and event-based models. Each approach reflects the components of terrorism risk (i.e., threat, vulnerability, and consequences) and their uncertainties in different ways. As examples of simple indicators, we describe how population and density-weighted population have been used as estimates of terrorism risk. As an example of event-based models, we describe the RMS Terrorism Risk Model. These two examples allow for comparisons that illustrate each approach's strengths and weaknesses.

Population is often incorporated into simple indicators of terrorism risk. The DHS State Homeland Security Grant Program uses a combination of equal allocation of resources and use of population as a simple proxy for terrorism risk. In its first several years of funding, the program allocated its first cut of funding so that each state received 0.75 percent of the total resources. The second cut allocated the remaining funding in proportion to each state's population. Proposed legislation would reduce the guaranteed amount directed to each state and require that the remaining funds are allocated with consideration of threat. The UASI grant program allocation formula, mentioned previously, is also partially based on population. There is a logical link between population-based indicators and terrorism risk.

An argument can be made that consequences are correlated with population. However, terrorism risks to a population of 100,000 are clearly different if that population resides in a dense urban area rather than if it is spread across a larger rural area because of the higher probability of many high-profile targets and more people within any given attack footprint. Density-weighted population, i.e., the product of a region's population and its population density, offers one of many possible simple risk indicators that account for this difference. Just as population can be considered correlated with consequences, so too is population density correlated with threat.

For example, a terrorist targeting 1,000 people might be more likely to attack a group when they are all within the same city block than if they are dispersed across the country. In evaluating alternative approaches for allocating resources, the Congressional Research Service noted that population and density-weighted population is correlated though result in different distributions of resources. There are practical benefits for using simple risk indicators such as those based upon population. In general, the metrics for measuring these indicators are well understood, measurable, and data is widely available. For example, the 2000 U.S. Census is a very credible source for data on the population and population density for the UASI funded urban areas.

However, there are limitations to using simple indicators of terrorism risk. The main limitation is that they do not fully reflect the interactions of threat, vulnerability. As a result, there is little consensus and no validated framework for deciding how to use several simple indicators to create a single risk estimator. For example, one simple indicator of risk with respect to biological attacks on livestock might be the size of the livestock population. However, there is no theoretical or empirical basis for deciding whether counts of livestock should be included as an agro terrorism indicator in a model of a region's overall risk, and if so, in what proportion to other indicators like population or energy infrastructure. Furthermore, including an indicator like livestock could dramatically alter conclusions about the distribution of risk across regions. In short, therefore, stakeholders with concerns about different types of terrorism are unlikely to agree upon any model of risk that relies on a single presumptive risk indicator.

Event-based models are built upon relatively detailed analysis of consequences from specific attack scenarios. These models include sensitivity analysis for important parameters that affect consequences. They may include components to model multiple types of events and multiple targets. They may also include modules that translate expert judgments of likelihood or consequences. The strength of event-based models lies in the greater detail they enable in analysis. The weakness is that to obtain this detail analysts must estimate many uncertain parameters that define the attack scenarios. One example of an event based model is the RMS Terrorism Risk Model.

The RMS model was developed as a tool for the insurance and reinsurance industries to assess risks of macro-terrorism. To reflect risk as a function of threat, vulnerability, and consequences, the RMS model calculates the expected annual consequences (human and economic) from diverse terrorist threats. The methodology relies on models of specific threat scenarios, calculations of economic and human life consequences of each scenario, and assessments of the relative probability of different types of attacks on different targets. The RMS model calculates the threat of different types of attacks at different targets using expert judgment about target selection by terrorists, capabilities for different attack modes, overall likelihood of attack, and propensity to stage multiple coordinated attacks.

The RMS model assesses vulnerability by taking into account how security measures may lower the overall level of threat to a class of targets or deflect risk from one target to another. It then calculates the consequences of terrorist attacks for 37 attack modes viewed as representative of the types of events that terrorists are capable of and motivated to attempt. Consequences are assessed in terms of economic losses, injuries, and fatalities using geocoded databases of population density, human activity patterns, business activities, and values of buildings and their contents. RMS selected the attack modes to be sufficiently distinct and well enough defined so that it is possible to specify scenario parameters so that the losses from the event can be modeled.

Sensitivity analysis is used to estimate the expected event outcomes given a range of relevant model parameters for each event scenario. For example, models using plume dispersion estimates take account of variation in wind speed, release point above ground, and atmospheric stability. However, the events were also selected to represent a sufficiently wide range of potential attacks to cover the loss potential from terrorism. While some may view this list to be comprehensive, others may notice that attacks involving suicide bombers, liquefied natural gas (LNG) tankers, or other commonly discussed attack modes have been omitted. This point identifies an important limitation of event-based models. That is, results are dependent on a large number of underlying assumptions.

The RMS Terrorism Risk Model allows analysts to test the robustness of model results to assumptions through parametric analysis of threat, vulnerability, and consequences. One way this is done is through use of elicited threat outlooks. The expert elicitation process used by RMS has produced three perspectives on terrorist threat for the next year: a standard, enhanced, and reduced threat outlook. All of these perspectives incorporate consideration of al Qaeda and associated groups, other foreign threat groups including Hizballah, and domestic terrorist groups. Each perspective represents an aggregation of different beliefs about terrorist motivations and capabilities.

The standard outlook assumes a terrorist threat that is primarily from al Qaeda, though potentially from other foreign organizations, with al Qaeda having a low likelihood of using CBRN weapons, and an overall low likelihood of multiple, coordinated attacks. The enhanced outlook reflects a greater likelihood that a terrorist attack might occur, that al Qaeda might use CBRN attacks, and that attacks would involve multiple, coordinated events. The enhanced outlook also has a higher Threat Reduction areas were allocated UASI funding in FY 2004, several of these were analyzed as larger urban areas because of how the RMS model is configured.

Factor, reflecting that with a heightened awareness of terrorism activities, security measures would be tighter and have a greater effect on reduction of secondary attacks after a single attack happens. Finally, the reduced outlook assumes that the only macro terrorist threats to the United States are those posed by al Qaeda. We used the RMS Terrorism Risk Model to calculate expected annual consequences of terrorist attacks (i.e., terrorism risk). Losses were expressed in terms of numbers of fatalities, numbers of injuries, and total property damage in dollars (buildings, building contents, and business interruption).

Impacts were then aggregated across the urban areas that received funding through the UASI grant program by summing the expected annual consequences for each of the attack mode target pairs modeled for an urban area. Specifically, Los Angeles and Long Beach, Santa Ana and Anaheim, and Minneapolis and St. Paul received separate allocations but were modeled as Los Angeles-Long Beach, Orange County, and Minneapolis-St. Paul, respectively. As a result, the analysis covers 47 urban areas instead of 50. Subsequent discussions only use risk estimates derived from expected fatalities. However, the data indicates that for results from the RMS model, expected fatalities, injuries, and economic losses are highly correlated with each other. For the purpose of estimating each region's overall risk, we must select a single risk estimate for each city.

Moreover, we would like the overall risk estimate to characterize risk well, regardless of which of the three threat perspectives proves to be closest to true risk. Doing so requires combining multiple estimates of terrorism risk. We calculated an aggregated risk estimator from the three sets of expected fatalities generated by the RMS model, c_{ij} 1, using a constrained optimization.

This functional form of objective function used in this analysis incorporates several assumptions. First, the objective function above minimizes underestimation error as opposed to overall error. This is based on our judgment that is better to minimize potential losses from terrorism that could result from underestimating risk than it is to ensure that each city has an equal chance of having its risk over- or underestimated. This assumption tends to favor regions that have a distribution of consequences with a tail representing very large consequences.

Thus, risk estimates for dense, urban areas will be larger using this approach than an ordinary sum of squares minimization. Increasing estimates for cities like New York means that the larger cities to do not bear a disproportionate share of potential risk underestimation. Another reasonable approach would be to minimize overall sum of squares with an objective function that set I_{ij} in the equation above to a constant value of 1. While this objective function is better understood mathematically, it is not more desirable normatively. The subsequent analysis only presents results using the objective function minimizing underestimation error. Analysis using an overall error objective function leads to qualitatively similar, though not identical results.

Second, using squared error assumes that larger errors are much worse than smaller errors. If underestimation error is linked to preventable fatalities or other damages, this assumption reflects realities about risk perceptions of catastrophic events. Finally, this analysis only accounts for risk as measured by expected fatalities. As discussed previously, a complete treatment of terrorism risk estimation must use methods from the literatures on multi objective decision-making to provide transparency into value judgments regarding balancing across different types of consequences.

Shares of total population across the UASI-funded urban areas are presented as filled circles. The size of city shares of risk using this measure ranges from a high of 0.076 of total risk (Los Angeles-Long Beach, CA) to a low of 0.004 (New Haven-Meriden, CT), with 14 metropolitan areas having shares greater than the equal-share line. Density-weighted population shares run from a high of 0.378 (New York) to 0.0003 (Las Vegas), thus resulting in a much larger spread of estimated shares of total risk than derived by the population estimator. Moreover, using density weighted population, just eight cities are found to have more than the equal-share allocation of terrorism risk.

Immediately apparent is that most of these estimates of city risk shares are several orders of magnitude lower than the population or density-weighted population estimates. The aggregated estimates range from 7.87E-8 (Pittsburgh) to 0.672 (New York), with just six cities having shares greater than the equal share. Interestingly, by more equally distributing underestimation risk, the aggregated estimator causes 32 of the UASI urban areas to be counted as having virtually no share of total terrorism risk. If, for instance, the $675 million FY04 UASI funds had been distributed in proportion to the aggregate estimate of city risk, these cities would have received less than $70,000 in total.

Most of the concentration in risk found with the aggregated estimator is due to similar concentrations in the RMS estimates of expected annual terrorism consequences for each metropolitan area. Across threat outlooks, New York has an average of 58 percent of total expected losses in the RMS model, and six of the metropolitan areas account for more than 95 percent of total losses. Clearly, the aggregated estimator is sensitive to this underlying model of terrorism risk. Alternatively, other terrorism risk models that reflect threat and vulnerability information not captured in the RMS model would be expected to yield an aggregated estimator that suggests less concentration of risk, if the model estimates were similarly less concentrated.

Shares of UASI funding closely track urban areas' shares of population. On average, city population shares differ from grant allocation shares by just 0.006, with the maximum discrepancy of 0.02 occurring for Los Angeles. If one believes the underlying assumptions of the RMS Terrorism Risk Model, then the distribution of resources does not match the distribution of terrorism risk. As stated previously, this might be acceptable because of issues of cost effectiveness of available risk reduction opportunities.

We have argued that uncertainties in the distribution of terrorism risk necessitate risk estimates that perform well across a range of assumptions about terrorist threat, vulnerability, and consequences. In this chapter, we describe a model for evaluating the performance of alternative risk estimates across a range of plausible terrorism futures. We use this model to compare the robustness to uncertainty of three estimates introduced earlier: the population, density-weighted population, and the proposed aggregated estimate as indicators of the share of terrorism risk for each of the UASI urban areas.

The Performance of Terrorism Risk Estimates

The population, density-weighted population, and aggregated estimates each offer a different solution to the problem of estimating cities' unobservable share of true risk, from here forward referred to as a city's risk share. Given these differences, it is important to know which indicator offers the most robust depiction of true risk. As we discussed previously, however, we are less concerned with the problem of indicators overestimating true risk than we are with underestimation of true risk; therefore, we define an estimator's performance in terms of how well it minimizes underestimation of each urban area's terrorism risk share.

Since true risk shares are not directly observable, we conduct a series of simulations to explore the performance of the estimators across a range of plausible values for each city's true risk share. We begin with a best-available estimate of urban areas' true risk shares, and then systematically allow simulated true risk to vary around these best estimates, examining how each of the risk-share estimators performs. In this way, we examine how robust the different estimators are to a wide range of plausible futures designed to represent uncertainties in terrorist motives, targets, methods, and capabilities that are fundamental to the description of cities' true risk shares.

Two sets of simulations were conducted. In the first, we took RMS estimates of expected annual consequences for each city as the best available estimate of true risk, and then systematically allow for the possibility that true risk may differ from the RMS estimates by up to two orders of magnitude. Thus, for instance, if the RMS estimate of expected annual terrorism fatalities for New York City is 304, we examine the performance of the estimators if true fatality risk for New York ranges from 3.04 to 30,400.

Other urban areas' true risk shares were simultaneously, and independently, allowed to fluctuate around their RMS risk estimates by up to two orders of magnitude. Because our aggregated estimator is derived from the same RMS estimates of city risk, this first simulation may lead to results biased in its favor. To address this limitation, we conducted a second simulation in which each estimator's performance was examined after assuming that cities' shares of true risk may vary by up to two orders of magnitude around their shares of total density-weighted population.

Across simulations, we also examine the largest such sum of squared error to establish the worst case performance of each risk-share estimator at each value of k. In addition to examining the performance of the three city risk share estimators for each of the two surrogates for true risk, we include a sixth estimator for comparison purposes. This random risk share estimator is recalculated in each simulation trial, and estimates each city's risk share as a random value drawn from a uniform distribution, $U[0,1]$, divided by the sum of the uniform values drawn for all cities in the current trial.

The simulations are designed to construct a range of plausible values for each of the UASI urban areas' shares of total terrorism risk, so that the robustness to different possible futures of the risk-share estimators can be compared. For this case, when true risk differs from the RMS estimates by up to one order of magnitude, share of total true risk for New York ranges from a minimum of 0.028 to a maximum of 0.961 of all UASI urban areas' total risk. Baton Rouge, in contrast, had a minimum true risk share of 0.000002, and a maximum of 0.0027 of all cities' risk. Similar variability in simulated true risk-shares was observed in each of the models. In the first series of models, simulated true risk varies around the RMS model estimates of expected fatalities, injuries, and property loss across terrorism risk outlooks.

As such, these models use all or a superset of the RMS model estimates of risk as the basis for simulating true risk (three types of consequences in each of three terrorism risk outlooks). Since the aggregated estimator was developed to minimize underestimation error using the RMS model, it might be expected to outperform the other estimators. Nevertheless, we include measurements of the performance of the aggregated estimator in the first series of models, because it provides information on how well an optimized risk-share estimator could perform, which aids in the interpretation of the performance of the other risk-share estimators. As expected, the random estimator is associated with the greatest underestimation error and the aggregated estimator is associated with the lowest underestimation error.

Interestingly, at all levels of k, the density-weighted population estimator resulted in underestimation error closer to that of the aggregated estimator than to that of the random or population estimators. The population estimator, in turn, had underestimation error that was consistently closer to that of the random estimator than to that of either the density-weighted population or robust estimators. As noted earlier, the simulation procedures allowed for considerable variability in simulated true risk estimates. For instance, in some trials New York accounted for less than one percent of total risk, whereas in others it accounted for more than 99 percent. Given this degree of uncertainty, it is useful to examine worst-case scenarios for the various city risk-share estimators.

In this worst-case analysis, the estimators exhibit the same rank ordering as found with mean underestimation error, with the random estimator performing most poorly and the aggregated estimator performing best. However, the maximum underestimation error increases only modestly for $k > 30$. These models test the performance of the risk-share estimators across alternative value judgments about what types of consequences are of greatest concern. Each of these models uses just a subset of the data used to calculate the robust estimator, and again it outperforms all other city risk-share estimators.

Interestingly, however, the aggregated estimator exhibits a comparable mean underestimation error to the density-weighted population estimator for higher levels of k. As in the first series of models, the population estimator produces underestimation errors closer to the random estimator than to either the density-weighted population estimator or the aggregated estimator. Here the aggregated estimator clearly exhibits higher underestimation error than the density-weighted population estimator, but otherwise the relative performance of the estimators is similar to what has been observed in all earlier models.

The consistency of findings across the models has several important implications. First, the aggregated estimator exhibits as low or lower mean underestimation error than all other estimators considered in this study. As discussed previously, this was expected in the first series of models in which the same data used to generate the robust estimator were used to simulate true city risk. Nevertheless, given that we cannot directly observe true risk, we must estimate relative risk based on some best-available model of city risk.

Our models show that using procedures like those we developed to produce the aggregated estimator, a one-dimensional estimator can be calculated that minimizes mean underestimation error across multidimensional model-based risk information, and which therefore exhibits lower mean underestimation error given uncertainties than alternative estimates of city risk shares. Similarly, even when worst-case scenarios are sought, defined as those simulated allocations of true city risk that maximize estimator error, the aggregated estimator exhibits lower underestimation error than all other estimators, or, in the case of Model 2, performed nearly as well as the best estimator. Second, the density-weighted population estimator results in substantially reduced underestimation error in comparison to the random and population estimators.

Moreover, in the first series of models, it performed nearly as well as the aggregated estimator, and in the second model, it was equivalent to or better than the aggregated estimator. The good performance of the density-weighted population estimator makes it particularly attractive for problems of establishing terrorism risk shares when sophisticated models of risk are unavailable or when there is a need for a simple and transparent risk model, as might be the case should such models need to be publicly debated. Third, across all the trials, allocating resources using the population- based approach fares little better than the random estimator that treats Baton Rouge as having a share of risk equivalent to New York—or flipping a coin.

The random estimator consistently performed worse than all other estimators, but perhaps not as differently as might be expected. This equal distribution of risk shares led to performance not far worse than the performance of the population-based approach. This finding raises important questions about the appropriateness of population as a risk indicator. Finally, the aggregated risk estimator's performance on different unidimensional risk perspectives is a measure of its robustness. In this study, the relative utility of the different city risk-share estimators does not depend strongly on whether or not terrorists are likely to employ weapons of mass destruction.

That is, when we base our simulation on true risk estimates that assume terrorists will not use weapons of mass destruction (the RMS Reduced terrorism outlook, we find essentially the same pattern of estimator performance as when we base the simulation on estimates of relative city risk that assume that the risk of a WMD attack is even greater than suggested by current conventional wisdom (the RMS Enhanced terrorism outlook). We find a similar invariance in estimator performance across consequence types (fatalities, injuries, and property loss). For the models used in this study, the aggregated estimator is relatively robust to fluctuations in the likelihood of WMD attacks and assumptions that prioritize lives or property.

In the discussion that follows, these observations are dependent upon the models used in this study and may differ if other models are considered. However, the approach used to assess robustness would remain valid. To summarize, the aggregated estimator performed well relative to other estimators, providing a proof of concept that a single, one dimensional estimate of city risk shares can offer a solution to the problem of determining cities' relative risks that can be robust to uncertainties about the likelihood of WMD attacks, uncertainties about which consequences ought to be prioritized, and uncertainties about the expected magnitude of risks each city might face. Furthermore, the density-weighted population estimator fared better than the other simple estimators and even outperformed the aggregated estimator in some cases.

Limitations of the Modeling Exercise

The findings in this chapter are subject to a number of model limitations, several of which we now consider. Clearly, the first series of models reported here depend on the assumption that the RMS estimates offer reasonably good approximations of the distribution of true terrorism risk across the 47 urban areas. Different assumptions about the distribution of true risk could lead to different conclusions about the performance of the different estimators. For instance, the RMS model describes terrorism risk largely as a threat to urban areas. Models that provide more detail into threats to rural areas, such as agro terrorism, would likely provide different observations. However, they, too, could easily be incorporated into the methodology demonstrated in this study.

By rerunning the model assuming that true risk is distributed proportionally to density-weighted population, we were able to offer a limited test of the sensitivity of our results to our reliance on RMS estimates. This tests a case where true risk is more accurately reflected with a simple model rather than a complex model like RMS. A more complete test would compare estimator performance when true risk is simulated using an independent event-based model of city risk. Our approach to simulating uncertainty in true risk shares and estimator performance relies on assumptions that may be incorrect or inadequate. In simulating true city risk shares, for instance, we alter each city's RMS risk estimate by a factor ku, where u is a random variable with a uniform distribution.

This approach ensured that approximately half of all trials would have simulated true risk values lower than the

RMS estimates, and half would have larger values. But alternative distributions for the random variable and alternative approaches to modeling variation in true risk are possible and might lead to different conclusions.

In each of these model series, the relative performance of the risk-share estimators for both mean underestimation error and maximum underestimation error was identical to those reported above when using a uniform distribution for u. Another modeling assumption we made was that the harms associated with underestimating city risk shares are not linearly related to the magnitude of the underestimation but rather grow exponentially with underestimation errors. As such, we measured estimator performance by summing the square of underestimation error, rather than, for instance, the sum of underestimation errors or other aggregations of error. This decision treats a risk underestimation error of 0.2 for some city as substantially worse than underestimation by 0.1 at two cities, for example. Although this assumption appears reasonable, the true relationship between underestimation errors and harms may be different and not well described by squared error.

A more complete test would compare estimator performance when error is assumed to be the absolute deviations instead of squared deviations. To improve the allocation of homeland security resources and thereby to reduce loss of life and property to terrorism or minimize poor investments in homeland security measures if attacks do not take place, it is essential to have good estimates of the terrorism risk to which different regions or groups are exposed. This objective has been difficult to achieve for many reasons, including confusion about the definition of risk and the absence of a systematic framework for selecting risk indicators. This monograph offers a definition of risk and discusses the relationships among threats, vulnerabilities, consequences, and risk.

In addition, it suggests a method for constructing a single dimensional estimate of city risk shares, designed to perform well across a wide range of threat scenarios, risk types, and other sources of uncertainty. Finally, it proposes and demonstrates a framework for comparing the performance of alternative risk estimates given uncertainty in terrorists' intentions and capabilities, target vulnerabilities, and the likely consequences of successful terrorist attacks.

Defining Terrorism Risk

Estimating or measuring terrorism risk is incomplete without a framework that considers threat, vulnerability, and consequences. To establish a specific and actionable framework for analysis we develop definitions for threats as external, dynamic forces acting on targets or infrastructure and vulnerabilities as properties of the targets themselves. Together threat and vulnerability define the probability that specific types of damage-causing attacks will occur at specific targets during specified periods.

The methods used to estimate threat are qualitatively different than those used to measure vulnerability and consequences. The former relies on collection and interpretation of intelligence. The latter requires scientific and engineering expertise of attack modes and target responses to attacks. Finally, threat, vulnerability, and consequences are interdependent since terrorism and homeland security is a multisided game, and terrorists may act strategically to increase their effectiveness. Thus, interactions between threat, vulnerability, and con-sequences have important consequences for risk management, though not explicitly discussed in these models.

Effective assessment of terrorism risks requires measures that can accommodate the uncertainty inherent in the problem. We have proposed a method for estimating terrorism risk that hedges against uncertainties in threats, vulnerabilities, and consequences. Similar to practices used in economic forecasting, this estimator aggregates information from multiple models or experts. The aggregated estimator is robust in that it reflects a range of assumptions about terrorist threat and perspectives on relative importance of different measures of consequences. Further, we generate this aggregated estimator using an approach that minimizes the extent to which risk is underestimated across urban areas. Since aggregation can mask important differences between models or experts, it is important to consider how this aggregated estimate differs from estimates based on single perspectives.

We describe and compare two approaches frequently used to estimate risk: simple indicators and event-based models. Examples of simple Conclusions and Recommendations 53 indicators include population and density-weighted population. We argue that event-based models offer a framework for calculating risk that overcomes some of the arbitrariness of simple indicators of risk that rely on presumptive correlate relationships. Supporting this claim requires an approach to evaluation that accommodates several types of uncertainty.

This method provides insights into both distinctions between risk estimators and the variance of single estimators to assumptions about threat, vulnerability, and consequences. This analysis showed that our aggregated estimator exhibited as low or lower underestimation error than risk estimates based on population and density-weighted population, demonstrating that a single estimate of city risk shares can offer a solution to the problem of determining cities' shares of total risk that is robust to a wide range of plausible terrorism risk futures. The density-weighted population estimator results in substantially reduced underestimation error in comparison to the random and population estimators. In many cases, density-weighted population performs comparably to the robust estimator.

This suggests that for some purposes, use of density-weighted population as a simple risk indicator might be preferred when, for example, they are informing decisions with a strategic time horizon and lead time. Decision makers may wish to estimate risk to inform strategic resource allocations, operational resource allocations, or to evaluate how terrorism risk is changing. Strategic allocations differ from operational allocations in terms of the frequency and ease with which allocation decisions can be changed. Homeland security resources might be allocated strategically when they are expensive or infeasible to move or change once committed. For example, resources directed toward training emergency responders are constrained by the time it takes to complete training and the resistance of emergency response personnel to frequent relocation.

On the other hand, operational allocations might be made in response to specific intelligence or to address short-term vulnerabilities. For example, the Democratic and Republican National Conventions in 2004 created new vulnerabilities, requiring enhanced security resources for a short period. Density-weighted population is more appropriate than either population or random estimators for informing strategic resource allocations. The underlying data are easily obtained and provide credible, face-valid indicators of risk. Both of these factors increase the utility of density-weighted population in public debates about resource allocations. Density-weighted population, however, does not allow decision makers to see how changes in threat or vulnerability information affect terrorism risk.

For example, when making operational resource allocations or evaluating the effectiveness of preparedness programs, decision makers need to understand how specific countermeasures reduce or change the profile of terrorism risk. Similarly, a crude indicator like density-weighted population would offer no guidance about how city risk estimates might change with, for instance, new intelligence about terrorist targeting or capabilities of using WMD attacks. For these purposes, more detailed event-based models of terrorism risk are essential. In this study, a single estimate was shown to be robust across uncertainties about the likelihood of WMD attacks, uncertainties about which consequences ought to be prioritized in considerations of city risk, and uncertainties about the expected magnitude of risks each city might face. For example, an important observation is that the risk profile of the 47 cities did not change significantly with the variability (or absence) of threats from weapons of mass destruction.

While the primary focus of this study was not to estimate precisely terrorism risk in the United States, this observation raises questions about the distribution over many areas. In particular, one question might be to investigate whether risk is a characteristic a region's infrastructure or population that is relatively stable across different threats. If so, this would be an important observation when it comes to policy and resource decisions. Observations about the variance of the aggregated risk estimate are dependent upon the results of the RMS model of terrorism risk and set of alternative estimates to which the aggregated estimate was compared. However, the proposed methods can be readily implemented with new data sources or other models. Adding new information to that provided by the RMS Terrorism Risk Model would presumably further improve the robust estimates.

The framework for defining terrorism risk and the analysis presented in this monograph, leads us to five recommendations for improving the allocation of homeland security resources.

1. DHS should consistently define terrorism risk in terms of expected annual consequences.

    Calculating expected annual consequences requires

accounting for threat, vulnerability, and consequences. Defining terrorism risk in these terms facilitates the incorporation of risk reduction as the goal of homeland security programs.

2. DHS should seek robust risk estimators that account for uncertainty about terrorism risk and variance in citizen values.

Given the tremendous uncertainties surrounding terrorism risk assessment, it is prudent to plan for the range of plausible futures that may play out. Several approaches are available for generating estimates of city risk shares that offer robust characterizations of risk across multiple uncertainties and perspectives on relative values of different consequences. Our approach to this problem ensures that underestimation error is minimized.

3. DHS should develop event-based models of terrorism risk, like that used by RAND and RMS.

Measuring and tracking levels of terrorism risk is an important component of homeland security policy. These data provide insight into how current programs are reducing risk and when and where new terrorist threats may be emerging. Only event-based models of terrorism risk provide insight into how changes in assumptions or actual levels of threat, vulnerability, and consequences affect risk levels. This characteristic is important for informing operational level problems such as deciding which security and preparedness programs to implement. Furthermore, event based models overcome a principal shortcoming of models that combine diverse risk indicators: They provide a coherent, defensible framework for selecting and combining information about threats, vulnerabilities, and consequences.

4. Until reliable event-based models are constructed, density weighted population should be preferred over population as a simple risk indicator.

Density-weighted population is reasonably correlated with the distribution of terrorism risk across the United States, as estimated by event-based models like the RMS Terrorism Risk Model. To support strategic policy decisions when the effects of new countermeasures or recent intelligence are not in question, density weighted population is a useful indicator in lieu of event-based models. In contrast, our results suggest that population offers a remarkably weak indicator of risk, not much superior to estimating risk shares at random.

5. DHS should fund research to bridge the gap between terrorism risk assessment and resource allocation policies that are cost effective.

Until the relationship between allocation amounts and risk reduction is understood, resource allocation decisions will not be optimized for reducing casualties and property loss. To these ends, DHS should evaluate the performance of the formula used to assign UASI grants using the approach presented in this study. Homeland security efforts will be greatly improved with better understanding of both the resources required to affect a range of countermeasures and the risk reduction achieved by affecting those countermeasures.

7

The Chinese and US News Coverage of Terrorism Abroad

This explores US and Chinese news coverage of four terrorist events that occurred in Spain and Russia between 2004 and 2005. These are the March 2004 bombing of a Madrid railroad station and the February 2005 car bombing near Madrid's convention center in Spain; the August 2004 bombing of two Russian airplanes shortly after leaving Moscow and the September 2004 attack on a southern Russian secondary school. They all resulted in many deaths or injuries. Four newspapers, The New York Times and The Washington Post in the United States, and People's Daily and Southern Urban Daily in China, are selected.

Adopting the method of content analysis, this study compares the coverage between the US and Chinese newspapers in 14 days following each event in terms of the quantity of coverage, the placement of the coverage, the type of news items, and the focus of story. This study shows that all these events, with the exception of the Spain car bombing event, which was minor compared with the other ones, received considerable coverage in both the US and Chinese newspapers. [China Media Research. 2006;2(2):74-84].

Terrorism has become a prominent topic for news media coverage. In the United States, terrorism became an especially important news topic following the destruction of the World Trade Centers in New York City and the attack on the Pentagon in 2001. As the international group Al-Qaeda claimed responsibility for both these attacks and the less destructive 1993 bombing of the World Trade Center, Americans became especially aware of potential threats from terrorists abroad. Thus the issue of international terrorism has become high on the news agenda in the United States. Unlike the case in the United States, in China terror activities were primarily confined to Xinjiang in the northwestern part of the country, where groups of separatists instigated bombings, assassinations and street fighting.

There was speculation that they might be receiving funding from outside of the country. In addition, since the 2001 attacks in the US by the Al- Qaeda group, China has cooperated with the international community in trying to combat terrorism. We assume that terror/terrorism may also now appear more on the Chinese news agenda. As a result, media coverage of terrorism has presented itself for academic studies in the field of journalism and communication in the world. This study sets out to compare the US and Chinese news coverage of four terrorist events. Considering the two countries' different experiences with terrorism, we focus on how Chinese and the US newspapers cover terror events that occurred in other countries.

This permits a comparison that is not affected by the domestic involvement of either country in the events studied. For this purpose, coverage of four terrorist events – two in Spain and two in Russia — are selected in this study. These are: (1) The bombing of two airplanes after leaving Moscow in August 2004; (2) the attack on a school in Beslan, southern Russia in September 2004; (3) the bombing at the Madrid train stations in March 2004; and the car bombing near Madrid convention center in February

2005. Specifically, this study evaluates the features of coverage of these four events, in terms of the quantity of coverage published, the placement of the coverage, the type of news items, and the focus of story.

Related Theory and Research News Values

Media coverage of international terrorism is closely associated with the concept of news values as criteria applied by media gatekeepers in news selection. In the United States, news values have long been discussed in the field of journalism and mass communication. For example, in 1922, Lippmann emphasized the importance of location and of the event being "obtruding above what is normal" in news selection "in his account of routinization of newsgathering." For another example, in 1940, Robert Park touched upon properties of the news report by comparing news with history. His findings incorporate the concepts of timeliness and unusualness/unexpectedness as among features of news that distinguish the news story from history.

For decades, textbooks on news reporting and writing have discussed the classic elements/factors that are regarded as news determinants. Factors most widely accepted as criteria for news selection include: timeliness or novelty, importance or impact, proximity, prominence or eminence, oddity or unusualness, conflict or controversy, and human interest. Among academic writings touching upon news values, some have studied news values in international/foreign news. Galtung and Huge's study (1965) of foreign news in the Norwegian Press was described as leading to "the first clear statement of the news values (or 'news factors') that influence selection."

They identified 12 conditions: frequency, threshold (including absolute intensity and intensity increase), unambiguity, meaningfulness (including cultural proximity and relevance), consonance (including predictability and demand), unexpectedness (including unpredictability and scarcity), continuity, composition, and reference to elite nations, reference to elite persons, personification/reference to persons and negativity / reference to something negative. To them, events "become news" to the extent that they "satisfy" the conditions. Shoemaker and her colleagues find that among the classic elements of news values, the geographical proximity of an international event is among the determining factors leading to whether or not it will be chosen. Similar finding has also been documented by a study on foreign news patterns by Stevenson and Cole.

They state that proximity and timeliness seem to be universal news values. Moreover, Shoemaker and her colleagues link the traditional news values with two constructs— "deviance" and "social significance" for evaluating the influence of biological and cultural evolution on news selection. If one uses this view to look at media coverage of terrorist events, the reason for the selection of such events is not hard to find: Such events, examined in this view, are deviant and also socially important as they often cause large casualties.

Furthermore, several scholars discuss news values in their studies on media coverage of terrorism. Weimann and Winn (1994) find that terrorist events can meet well the 12 conditions of media coverage proposed by Galtung and Ruge. Hoge (1982) states, "Why cover terrorism? The reason is that terrorism is news." He holds that terrorism is "different, dramatic, and potentially violent." And "it frequently develops over a period of time, occurs in exotic locations, offers a clear confrontation, involves bizarre characters, and is politically noteworthy" and, "it is of concern of the public."

In China news values have been discussed since the period of the May 4th Movement in 1919. In the first book on journalism ever published in China, Xu, Baoheng, one of the forerunners of journalism educators, mentioned significance/importance, novelty/freshness and proximity as elements of news values. He writes, "News values involve the number of people who pay attention to the news event and the degree of intensity of their attention to the news event. Significant recent events naturally attract the attention of large numbers of people and lead to great intensity of attention. ... So such news events have high news values. ... We can thus work out a formula, i.e., the news value of an event is directly proportional to its significance/importance, in other words, it is directly proportional to the number of people paying attention to and the intensity of their attention."

In contemporary China, the concept of news values surface in a number of studies. For example, Tong (2004) summarizes elements of news values into five categories: novelty/freshness/timeliness, significance /importance, prominence, proximity and interest. Hu (1995) and Huang (1995) also list novelty/freshness/timeliness, significance/ importance, proximity, prominence and interest as generally accepted news values. Trying to link the characteristics of news items with audience's needs, Liu, Weidong (1999) points out that news values reflect not only the peculiarities of the facts, but also the degree to which they meet the audience's information needs. He thus suggests eight criteria for evaluating newsworthiness.

Liu, Jianming (2002) also emphasizes meeting audience's needs, holding that "The so-called modern news value is the effect and significance a piece of news shows when it meets the audience's needs. ...Usefulness, benefits and effectiveness are the essential elements of news values." We can find a fair similarity between the Chinese scholars and foreign scholars in their views regarding the basic news values. The absence of "oddity"/"unusualness" in the news values listed by these Chinese scholars is perhaps the only clear difference. Although studies touching upon news values are not rare in China, important empirical studies examining newsworthiness in media coverage seem to be a rarity in the country.

Thus this study chooses to adopt the empirical methodology in examining the US and Chinese news coverage of the four foreign events and exploring newsworthiness as reflected in the coverage.

Media Coverage of Terrorism

As frequency and intensity of terrorist activities catch much media attention, media coverage of terrorism has lent itself to academic studies by communication scholars in the United States in the past two decades. Among a number of studies that focus on how mass media cover the terror acts and actors Besides, several studies look into how media label terrorist actions and perpetrators; other studies review reporting formats and story types; still others observe isolate events.

Moreover, there are also some studies, fewer though, that focus on specific terrorist organization, e.g., IRA. Picard finds four patterns regarding the types of terrorism getting coverage: (1) state terrorism is generally ignored; (2) the amount of non-state terrorism covered is limited; (3) news coverage focuses on incidents and government issues at the expense of background information, and (4) there is a similarity in what gets covered among media. He also points out that the "predominant coverage emphasizes action, violence and government view, and provides little information that increases public understanding". He also concludes that not much variation can be found in "what is reported and how it is reported" by major American media because "they operate similarly and follow standard professional norms."

However, he admits that compared with the general coverage, coverage about specific events occasionally become overwhelming during a certain period. Regardless of numerous studies by the US communication research community on media coverage of terrorism, very few studies have been carried out to investigate how media in different countries cover a same terrorist event. In China, articles that comment on international media coverage of terrorism are scattered in journalism and communication journals, such as Fudan Journalism Studies Quarterly, Journal of International Communication, Chinese Journalists and so on, in newspapers, and on such media websites as the People's Daily's and Xinhua News Agency's or websites that dedicate to journalism and communication studies, like mediachina.net.

Such are focused on such issues as problems in media coverage of terrorism, role of media in advocating anti-terrorism efforts, challenges brought by the use of media by terrorists and so on. A recent article by Xu, Bing and Zhang Changjun, for example, discusses how media can serve as the carrier for launching anti-terrorism campaigns.

Research Questions and Hypotheses

China and the United States differ in many aspects: in social environments, in linguistic habits, in patterns of media presentation and, most importantly to this study, in their recent experience with terrorism. Our project therefore tries to test whether differences between China and the United States affect their coverage of these terrorist events occurring in the international arena. In this paper, we mainly pay attention to testing whether there are differences in news coverage of the four events in terms of quantity, i.e., how much coverage the events received, the placement of the coverage, which implies the prominence of the coverage, and the focus of the story (which suggests the main underlying emphasis of the news item), and in the types of news stories covering the events.

Quantity of the coverage is further examined in two aspects: the number of stories and the length. The research questions and hypotheses of this paper are: Research Question One: Is there a difference in the number of the stories covering the terrorist events abroad between US and Chinese newspapers? Research Question Two: Is there a difference in the length of stories between the two countries' newspapers and between the two countries' events? Is there an interaction between the country of newspapers and the events on the length of stories?

Hypothesis 2a: US newspaper stories about the events were longer than Chinese newspaper stories.

Hypothesis 2b: Stories about Russian events were longer than stories about Spanish stories.

Hypothesis 2c: There was an interaction between the country of newspapers and the events on the length of the stories. Research Question Three: Is there a difference in the placement/position of the stories covering the four terror events abroad between the US and Chinese newspapers?

Hypothesis 3a: There was a difference between the two countries' newspapers in placement of the coverage.

Hypothesis 3b: The US newspapers positioned coverage about these events more prominently than Chinese newspapers. Research Question Four: Is there a difference in focuses of the news stories covering the four events between the US and Chinese newspapers?

Hypothesis 4: In the US newspaper coverage, focuses of the news stories were likely to be diversified, whereas in the Chinese newspaper coverage, the news stories were most likely to be focused on the process/course of the event.

Hypothesis 5a: There was a difference in the types of stories that covered the four terror events between the US newspapers and Chinese newspapers; US coverage included both news and opinion-type articles, whereas the Chinese coverage mostly emphasized news.

Hypothesis 5b: There was a difference in the item types between Russian event stories and Spanish event stories in the US and Chinese newspapers respectively.

Methodology Data Sources

Two newspapers are examined in each country, The People's Daily and The Southern Urban Daily in China, and The New York Times and The Washington Post in the United States. Of the two Chinese newspapers, the People's Daily is the most authoritative newspaper representing the party in China. For "policy information and resolutions of the Chinese government and major domestic news and international news releases from China," the People's Daily clearly is the most influential newspaper in China. The Southern Urban Daily is owned and run by the Nanfang Press Group, a media corporation in China's Guangdong Province, with Nanfang Daily, the famous party paper in Guangdong, being the corporation's flagship paper.

As the first mass medium that disclosed the SARS within China in winter 2002, it rapidly built up its reputation afterwards, both nationally and internationally. As their counterparts in the US, The New York Times, as a leading newspaper, has been analyzed in numerous studies and involved in almost all studies about the coverage of terrorism. The Washington Post, carrying a lot of information about the federal government and foreign affairs, has been investigated in various studies on foreign news, some of them discuss coverage of terrorism.

Because these events occurred outside of the countries we are studying, we assume that coverage will drop off quickly over time – timeliness and proximity being important news values. In addition, there may be a (time) lag between the event and the beginning of its news coverage because of newspaper schedules or news priorities. Thus we assume that the most coverage of the events will occur within seven days (one week) of each event's occurrence. However, there may be additional stories in the week following that period, and the quality of these later reports may differ from earlier ones, perhaps being more analytical or, in the case of the Madrid train bombing, revealing that the Basque separatist group, that was speculated to be responsible by initial reports, was in fact not the responsible party.

So we chose 14 days following each event: (1) August 25 to September 7, 2004, for airplanes bombing in Moscow; (2) September 2-16, 2004, for the school hostage crisis in Russia; (3) March 12-25, 2004, for the bombing in train stations in Madrid; and (4) February 10 -23, 2005 for car bombing in Madrid.

Variables

The news item, defined in general as a set of contiguous verbal content elements is used as the unit of analysis. Visuals are not included in this study. In this paper, the following four variables are discussed.

1. Focus of story is a variable to measure the primary underlying emphasis or central meaning of the news item. Four categories of focuses are identified: the process or course of the event; current political climate or context around the event; historical background information and other terrorism unrelated to the event.

2. Item length calculates the word account of textual elements in newspapers. Specifically, we estimate the total number of separate English words or Chinese characters in each item, using Arabic numerals to calculate, excluding both space between words and punctuations, but headlines are included.

3. Type of item measures whether the news item emphasized only factual information, or made an analysis or conjecture about the event's causes. This study includes six types of news items: straight or hard news, soft news, editorials, commentaries, columns and letters to the editor.

4. Item position refers to the page on which the news item was placed within a specific newspaper, that is, the front page, the section front page and any other position.

Data Collection and Analysis

Coding of variables was done in each country to ensure that the cultural context of the data is maintained. Three Chinese master students coded Chinese newspapers at a university in Shanghai. Two master students in a communication school at a private university in the US coded US newspapers using the Lexis-Nexis database. Intercoder reliability is according to Scotti's phi for US newspapers. This study finds that coverage of the four terrorist events abroad in the four US and Chinese newspapers in the period of 14 days following each event amounted to 271 items, with the two US newspapers publishing 129 items and the two Chinese newspapers publishing 142 items.

For Research Question One, data from this study show that the two Chinese newspapers published more stories about the terrorist events than the two US newspapers. Additionally, there is a difference in the number of the stories covering the two countries' terrorist events between the US and Chinese newspapers. Among Chinese newspaper stories, more than two thirds of them (69.7%) covered Russian events while only less than one third (30.3%) covered Spanish events. For the US newspapers, the gap in the number of stories covering the four events in the two countries of Russia and Spain is not so big.

Furthermore, when we look at the number of stories for each of the four events respectively, we find that the rankings of the amount of coverage of the events are the same in the US and Chinese newspapers. Among the four events, the event that received most coverage in the US newspapers, i.e., the September 2004 attack on a southern Russian secondary school, also got most coverage in the Chinese newspapers, and the event that received least coverage in the US newspapers, i.e., the February 2005 car bombing near Madrid's convention center in Spain, also got least coverage in the Chinese newspapers.

With regard to Research Question Two, Hypothesis 2a (US newspaper stories were longer than Chinese newspaper stories) and Hypothesis 2b (Stories about Russian events were longer than stories about Spanish stories) are supported by the findings of data analysis, but Hypothesis 2c (There was an interaction between the country of newspapers and events on the length of the stories) is not supported. Regarding Research Question Three, data analysis shows that there is a difference in the placement/position of the stories covering the four terror events abroad between the US and Chinese newspapers, supporting Hypothesis 3a, but it does not show a simple support or non-support to Hypothesis 3b (The US newspapers positioned coverage about these events more prominently than the Chinese newspapers).

As regards Research Question Four, data indicate that although focuses of the news stories covering the four terrorist events abroad in the US newspapers are more diversified than in the Chinese newspapers, the difference is not big, as focuses of the stories on these events are mostly concentrated on the three categories of "the process/course of the event primarily," "current political environment or context of the event," and "historical background information about the event" in both the US newspapers (91.47 percent) and Chinese newspapers (97.18 percent); for other types of focuses of the news stories in the US newspapers, the story.

Omission is made here in the bar chart when the story numbers of the two countries with a certain type of focus are 0 and 1 respectively.) Regarding Research Question Five, data analysis indicates that the difference in the types of stories that covered terror events between US newspapers and Chinese newspapers lies in more opinion-type articles being included in the US coverage than in the Chinese coverage. Hypothesis 5b, i.e., there was a difference in the item types between Russian event stories and Spanish event stories in the US and Chinese newspapers respectively, is not statistically supported. This study selects four leading newspapers in China and the US to examine their coverage on international terrorism. The primary concern is paid to their quantitative attributes, in terms of four aspects — the quantity and placement of the coverage, the focus of the news story and the type of the news item.

Data analysis demonstrates that all these events, with the exception of the Spain car event, which was minor compared with the other ones, received considerable coverage in both the US and Chinese newspapers. This suggests that these events were covered in both countries due to their newsworthiness. Theses events that received considerable news coverage involved lives or safety of a large number of people, and as such they were socially important. They were also deviant or unusual in that they deviated from the norm. In addition, they also exhibited timeliness. Geographic proximity, however, does not seem to assume a clear role in affecting either the quantity of the coverage or the placement of the coverage. It can also be found that although there is a difference in the news coverage of the four events between the US and Chinese newspapers it is more subtle and complex than what is assumed by the hypotheses.

Regardless of the common impression that the United States has put the issue of terrorism very high on the media agenda and public agenda, the two US newspapers published fewer stories covering the four terrorist events abroad than the two Chinese newspapers. One approach to seeking an answer will be to link it to China's opening towards the outside world, which has stimulated increased media attention and public attention to important international news events. In particular, Guangzhou, where one of the two Chinese newspapers chosen for study is published, is among those coastal cities that took the lead in opening towards the outside world when China started its economic reform since the late 1970s, there appears to be a good reason to assume that in this city, media publish more foreign news than their peers in other parts of China.

Data show that US newspapers published longer stories on the average than that in Chinese newspapers. To find the answer to the underlying reason of the difference, however, further studies are needed. Data analysis, rather than giving a clear-cut support to the hypothesis of the US newspapers placing the coverage of these events more prominently than the Chinese newspapers, illustrates that the difference in the placement of news coverage of the events between the two countries' newspapers is mainly reflected in twice as many front page stories but one-eighth of section front page stories (5 vs. 40) in the US newspapers than the Chinese newspapers.

And, the number of stories placed on any other pages is exactly the same between the US news coverage and Chinese news coverage. A closer look at the data gives us an idea about the possible reason for that. The US newspapers place 25 stories on the front page for the Russian school event, and 17 stories on the front page for the Spanish train station event, but only 3 front page stories in the coverage can be found for the Russian airplane event and one for the Spanish car event.

The Spanish car event did not cause much loss of lives, and so this may be the reason why the coverage on this event is not prominently placed in the US newspapers. But this explanation does not provide an answer to the US newspapers' placement of the coverage of the Russian airplane event as this event caused big casualties. For the Chinese newspapers, however, the data seem to support a linkage between the event characteristics and the placement of the news coverage.

The Russian airplane event and Spanish train station event, both causing big casualties, each received front page coverage of 8 stories and more section front coverage; the

Russian school event, which threatened the safety of school kids, received front page coverage of 5 stories and more section front coverage. In contrast, the Spanish car event, a minor one compared with the other three events, received neither front-page coverage nor section-front-page coverage.

According to the content analysis, although there is a difference in the focus of the item between the US news coverage and the Chinese one, the difference is negligible in the sense that apart from the three categories of "the process/course of the event primarily," "current political environment or context of the event," and "historical background information about the event" that together account for the over 90 percent of the focus in both the US newspapers (91.47 percent) and Chinese newspapers (97.18 percent) and that other types of focuses amount just to one story each with the exception of Type/Category 5 (suspect) (2 stories) and Type/Category 6 (impact on EU) (3 stories).

These results may indicate that event process, political environment and historical background information present logical focuses for covering international events of political nature in both countries. Finally, in this study we find that the two US newspapers include more opinion-type articles than the two Chinese newspapers. We assume that the characteristics of the two US newspapers chosen may be related to this: As elite or quality newspapers, The New York Times, which has traditionally provided opinion in international events, and The Washington Post, famous for interpretative reporting, seems to tend to interpret when covering many international events.

However, we admit that content analysis alone is not enough for us to draw a definite conclusion about the underlying reason for the story types in the US news coverage of the four events. By comparison, the two Chinese newspapers paid more attention to hard news when they cover international terrorism. In summary, although data analysis of this study indicates a difference in the news coverage of the four events between the US and Chinese newspapers, further studies are needed for detailed analyses on the underlying reasons of these differences.

8

ASEAN Counter-terrorism Treaty

The benefits for Association of Southeast Asian Nations (ASEAN) members of a regional treaty to combat terrorism include improved coordination in mutual legal assistance and harmonisation of best practice legal approaches. The conceptual framework for a common definition of terrorism is set out in this paper. Precedent regional and multilateral treaties are analysed into legal formulae and their components, such as obligations to indict or to extradite, to provide mutual legal assistance, and to build regional implementation capacity, are assessed as potential models for inclusion in an ASEAN regional treaty. The paper concludes by considering ASEAN progress in adopting cooperative mechanisms to combat terrorism thus far.

What might be the benefit of a regional counter-terrorism treaty to ASEAN Member States? The answer lies in modern terrorism's transnational dimension. It typically involves criminal actions across national borders that require international cooperation to combat them. Harmonisation of relevant national laws across those borders can facilitate enforcement cooperation and best national practice. The objective of this paper is to consider the elements that should be included in an ASEAN regional treaty to combat terrorism. Examples of such elements are:

national adoption of common terms and definitions, obligations to indict or to extradite, and reciprocal provisions governing mutual legal assistance and punishment.

It examines ways a regional treaty might promote harmonised definition, application and enforcement of national counter-terrorism laws. Ideally, the treaty should promote its own implementation by building regional capacity to enforce national laws through cooperative intelligence gathering, prevention strategies, crisis management and investigation efforts. This study commences with an examination of the conceptual issues, focusing on the legal definition of terrorism. This is an essential first step to establishing a common legal framework.

It then surveys concepts, standards and mechanisms utilised in other global and regional treaties. These are compared as model elements for adaptation into an ASEAN regional treaty. They are assessed according to whether they would function consistently with the adopted conceptual framework and are self-evidently clear and fair. The paper concludes with observations on current steps, as at June 2004, towards the possible development of an ASEAN regional counter-terrorism convention.

Conceptual Framework

A definition of terrorism in international law is necessary to provide a clear reference point for triggering and limiting international obligations to cooperate across a broad range of relevant law enforcement, intelligence-gathering, incident prevention and emergency response measures. Yet, for decades, international efforts to formulate a global definition for terrorism have been mired in political and ideological controversy. It is argued here that the mire is not deep and can be safely crossed. The controversy often begins with the argument that the concept of terrorism is an arbitrary exercise in pejorative labelling of certain acts that are not essentially different from other acts of political violence, such as the conduct of war, or of violent crimes such as kidnapping.

The cliché, that "one man's terrorist is another man's freedom fighter", takes as its premise that evaluations of the legality of political violence manifest subjective perspectives concerning its political motivation. This position is used to provide impunity from judgement of all types of non-military acts of political violence, rendering them all, including atrocities, equally acceptable, irrespective of circumstance or manner of conduct. The criminality of non-military political violence becomes a matter of subjective and arbitrary opinion.

In denying judgement of non-military political violence, this pluralistic relativism undermines the international and domestic legal orders that protect public security. Instead, we posit a definition of non-military political violence that presumes that the violence is motivated by a political, religious or ideological objective but eschews any evaluation of the objective itself.

Definition of Terrorist Acts

The international legal definition of terrorism proposed here avoids evaluation of any ideological purpose. Instead, it is concerned with the qualities of a "terrorist act". The four qualities of the act are: (a) serious violence; (b) intended to influence a public or its institutions; (c) by intimidating civilians in that society; and (d) committed by non-State actors. The basic element of a terrorist act is serious violence. Examples of the sorts of actions include murder, kidnapping, assault, grievous bodily harm, hostage taking, hijacking, malicious damage to property and major interference with communications. This basic element in the definition, *i.e.*, the deliberate perpetration of a violent action or part thereof, stands irrespective of whether the particular act has already been criminalised by the State under one of the above categories of violence.

In fact, the acts listed above are usually already criminalised, although this is not the case with all related preparatory acts (*e.g.*, planning, intelligence and equipment gathering, recruiting, training, financing, threatening, *etc.*). Although the related preparations are essential aspects of the violent action, they often may not yet be criminalised, as many legislatures have not anticipated the organisational sophistication that frequently characterises conspiracies to commit terrorist violence. The second conceptual element in the definition is strategic motivation. The violent action must be a step in an on-going campaign intended to achieve a political, ideological or religious objective.

Thus, terrorism is distinct from ordinary violent crime committed by private persons because its motivation is societal change. Rather than the private financial gain or emotional satisfaction that motivates violent crimes of a private nature, a terrorist act has a public motivation. The action is, therefore, distinguishable from common violent crimes, such as murder, robbery and vandalism, which are generally committed for private benefit. This public quality, of attack on social and governmental institutions, gives terrorist acts some of the character of armed conflict. Military operations, revolutionary and guerrilla conflicts also involve violence as part of a strategic campaign. However, terrorist campaigns cannot be subsumed into the category of legitimate armed conflict. Because they target civilians, they will always remain outside the norms of legitimate armed conflict that seek to protect non-combatants.

The third element of the definition is the intention to harm civilians. In the twenty-first century, grotesque violence targeted to maximise civilian deaths and injuries are deliberately publicised with the intention of terrorising the public. The victims of terrorist acts not only include the dead and wounded but terrorised witnesses too. Widespread public terror is strategically intended to coerce governments, as indicated above. "Civilians" can be broadly interpreted to include all "non-combatants" although the latter term is vexed by questions of interpretation of its scope in international law.

Therefore, for the sake of simplicity, the more limited term "civilians" is used here. Exclusive applicability to persons that are non-State actors, or are not ostensibly acting in an official State capacity, forms the fourth element. State actors are already bound by applicable international laws that circumscribe the legitimate use of political violence and categorise its breach as war crimes or crimes against humanity or as breaches of State responsibility (these are dealt with in more detail below under "State actors"). Unfortunately, international laws do not currently specify objective or appropriate rules for comparable actions committed by non-State actors who flout accepted humanitarian norms.

It does not make sense to redesignate established categories of the State's illegal use of political violence as terrorism while the current epidemic of political violence initiated and perpetrated by non- State actors continues to be neglected under international law. Therefore, a definition of terrorism that is legally useful must address political violence conducted by persons who are not State actors. In summary, a terrorist act is different from both common violent crime and armed conflict because it embodies qualities of both and is exclusively neither. The violent crime cannot be subsumed into ordinary criminal categories because its objective is to subvert political processes and it cannot subsumed into the category of armed conflict because it is entirely criminal in method. Violent Act Common Crime non-State actors civilian targets Armed Conflict political goal fear strategy.

Non-State actors

The dual character of terrorism, as violent crime and armed conflict, was recognised by the United Nations

Security Council in its Resolution 1373. It described the 11 September 2001 bombings in the USA as international terrorism and expressly connected them with both the right to self-defence (preamble) and with transnational organised crime (para. 4). Thus, it can be asserted that non-State actors can perpetrate international violence on a scale sufficient to amount to an armed attack that triggers the target State's right of selfdefence.

The Foreign Ministers of Security Council Member States made the connection once again in Resolution 1456, by describing terrorism both as criminal action and as a most serious threat to international peace and security. Nevertheless, partly because the primary objects of the international legal system are States rather than non-State actors, there is no international law defining and criminalising terrorism. As demonstrated below, even international humanitarian laws that generate individual criminal responsibility do not define or criminalise terrorism. Just as a regular State army can manage legitimate military operations or deviate into war crimes, a private non-State fighting organisation can discipline itself to conduct legitimate military operations or it can opt to commit crimes.

Non-State guerrilla fighters are recognised as legitimate international combatants in certain circumstances under the 1949 Geneva Conventions and 1977 Protocol I. Internal discipline to ensure their compliance with international humanitarian law is, arguably, a precondition to the fighting organisation being recognised as a legitimate armed force, although breaches by its individual combatants need not negate the organisation's armed forces status. The organisation's acts of international guerrilla warfare against proper military targets conducted in accordance with internationally allowed methods would, therefore, not breach international humanitarian law. The legal distinction between legitimate guerrilla warfare and terrorist acts by non-State actors is one of methods and targets.

All political violence in the form of deliberate attacks on civilians by non-State actors would be criminalised under international humanitarian law, if the attackers were subject to it. Unfortunately, however, non-State actors are outside the obligations of international humanitarian law in a wide range of circumstances. For example, combatants who are not regular armed forces or assimilated therein are not obligated under the Geneva Conventions. Similarly, liberation organisations are not "High Contracting Parties" to the Geneva Conventions and Protocols, and although they may expressly choose to abide by the norms of armed conflict, they need not do so.

Only within the relatively narrow range of "crimes against humanity", as described in customary international law and within the Statute of the International Criminal Court, are non-State actors criminalised under generally applicable international humanitarian law for deliberate attacks on civilian targets. The wider range of circumstances in which political violence against civilians is perpetrated by non-State actors to intimidate a public for political purposes is not addressed. Under international humanitarian law, the cohorts of non-State fighters deliberately attacking civilians are not declared *"hostes humanis generis"*, but are instead afforded the conventional rights, including "prisoner of war" or "protected person" status.

This nonreciprocal approach to obligations in international conflict was designed during the early post-colonial era to benefit various ongoing struggles for self-determination. Significant continuing international support for unconstrained political violence in pursuit (by "all available means") of undefined "peoples' struggles" suggests that targeting civilians still remains politically acceptable to many States at the commencement of the 21st century. The failure to agree on terms to criminalise terrorist activity under a generally applicable international law is one of the most regrettable derelictions of the United

Nations General Assembly's history and the Red Cross Movement has done no better.

International laws already identify a wide range of other crimes that can be committed by non-State actors, from drug trafficking to people smuggling, and customary law addresses crimes from piracy to crimes against humanity. The time has come to close the many gaps in international law concerning non-State acts of political violence against civilians. A sophisticated terrorist act requires planning, intelligence, financing, equipment, technology, publicising, training, political support and a frontline. These are diverse roles distributed among actors who are precisely co-ordinated together. In most instances, therefore, sophisticated or on-going terrorist actions require a coordinating organisation and, often, support from benefactor States.

A multi-faceted and complex organisation might engage in various acts, some bearing the characteristics of crime, others of armed conflict, others of a legitimate political nature. Thus, an organisation might engage in terror acts, while also being dedicated to a range of other social functions that serve, in part, as a cloak for foreign State support. The HAMAS organisation undertakes political and welfare activity but also perpetrates terrorist acts through its paramilitary forces, *i.e.*, the Izz-Al- Din-al-Qassam Brigades. Drug trafficking by the FARC in Colombia might occur simply as violent crime rather than fundraising for a terrorist act.

An organisation's personality is every bit as complicated as a natural person's. Thus, a terrorist organisation can also be a political, military or criminal one and the qualities are not mutually exclusive. The involvement of States in the activities of terrorist organisations varies widely. Involvement may be directive or at arms' length and may provide support across logistical, intelligence, financial, procurement, training, communications, refuge, advocacy and other needs. In contrast, full State control over political violence against civilians entails direct responsibility and more severe legal consequences, as discussed below. Non-State Actor Violent crime Terrorist act War crime (if under Geneva conventions) Guerrilla warfare (if under Geneva conventions).

Although some political commentators describe certain States as "terrorist States", this populist rhetoric obfuscates, rather than assists, legal analysis. Unfortunately, some senior legal commentators presume, by casual reference, the notion of "State terrorism" without any supporting legal argument at all. The possibility of generating any criminal responsibility of States in international law has often been discussed, but remains speculative. International law has already formulated other legal categories to characterise political violence against civilians when perpetrated by individuals acting directly on behalf of States. These are characterised as war crimes, as crimes against humanity, or as breaches of State responsibility. Each of these is surveyed briefly below.

First, acts in armed conflict are subject to the international laws of armed conflict. When an armed attack or civil war does take place, State parties to the conflict become subject to the applicable laws of armed conflict and breaches of those laws then constitute war crimes. Legal uncertainty persists over when foreign State intervention might amount to an illegal international armed attack. Thus, a cross-border foray by an armed band of irregular fighters can be considered an armed attack, but not the cross-border provision of weapons or logistical support to local rebels.

Domestic rioting that occurs on a scale large enough to be considered civil war is also subject to the laws of armed conflict but, again, there is no legal clarity as to when civil disturbance reaches the point of civil war. Thus, the circumstantial point at which attacks by States on civilians can be categorised as war crimes, remains to be assessed in each instance. Second, international laws concerning genocide and 'crimes against humanity' impose culpability upon State actors for widespread or systematic crimes committed against civilian populations, both within and outside of the bounds of an armed conflict. The rules are designed to address acts of a State, such as against its own or foreign citizens. State involvement—at least approval—is essential to this category of crime. Thus, crimes against humanity include the actions of non-State actors if they act with the approval of the State. Third, the international legal doctrine of State responsibility renders the State generally liable for acts in breach of its international obligations.

In the context of illegal use of force, the breach of international obligation attributed to the State can take the forms of overt armed State attack, or covert State attack, or attack conducted by non-State proxies. Examples in each respective case are the national responsibility of Iraq for its armed attack on Kuwait, of France for its bombing of a vessel in New Zealand waters, and of the USA for its support for insurgents in Nicaragua. State support for political violence by non-State actors against civilians is usually directed against a foreign public and undertaken at arms' length, using covert State agents or proxy organisations so as to avoid the grave consequences of overt armed conflict or of international responsibility.

Examples of the use of covert agents against foreign civilian targets include the Lockerbie and *Rainbow Warrior* bombings. Examples of the use of proxy organisations to attack civilian targets include Syrian sponsorship of Palestinian terrorist organisations and Iran's sponsorship of Hizbullah. When the veil of deception is pierced to reveal that a State actively supports terrorist acts perpetrated by private individuals against another State, legal consequences arise. The emerging practice is to hold responsible both the State, through international law, and the individual, through national criminal law.

Most notoriously, the 1988 bombing of Pan Am flight

103 over Lockerbie, Scotland, was categorised as both a private act and as an act of State, *i.e.*, as a transnational crime by a private individual under Scottish law and as a foreign State attack in breach of international law. Consequently, the Lockerbie bombing attracted both a criminal penalty for the individual Libyan officer convicted (a prison sentence served in The Hague by Mr. Meghrahi) and State responsibility (UN Security Council sanctions and compensation requirements imposed upon Libya).

Similarly, Taliban (and Al Qaeda) fighters were held responsible individually, together with Afghanistan, for the 11 September 2001 bombings in the USA. In the case of the *Rainbow Warrior*, a settlement was reached that imposed penal sentences under New Zealand criminal law on the French perpetrators, Alain Mafart and Dominique Prieur, and US$7 million compensation and apology obligations on France. In each case, terrorism is properly treated as a crime committed by individuals to be punished under national laws, whereas State involvement in hostile acts is treated in international law as an illegal armed attack or as a breach of state responsibility.

These dual approaches are consistent with the dual approach of the UN Security Council, described above. Ultimately, where a State is involved in the commission of an international terrorist act, whether by its clandestine officials or through a sponsored organisation, the State's action is appropriately addressed under international laws of armed conflict, crimes against humanity or State responsibility. Conversely, the individual persons concerned are culpable under criminal law.

The legal personality of the perpetrator is relevant to determine the legal system applicable to it. Both national judicial organs (*e.g.*, national tribunals exercising universal jurisdiction) and international judicial organs (*e.g.*, the International Criminal Court) can exercise enforcement jurisdiction over individual persons accused of committing internationally proscribed crimes such as genocide and "crimes against humanity". However, there are far more national tribunals and it is more usual that they will be mandated to exercise jurisdiction.

The legal definition of "terrorist acts" proposed here has not yet crystallised as an internationally proscribed crime and, as yet, there are no international tribunals mandated with jurisdiction over it. Its adoption in international law could serve two functions: to promote its wider incorporation into national laws; and, eventually, to form a basis for the exercise of criminal jurisdiction by an international judicial organ.

At present, it can be readily applied in national laws because its objects are individual persons who are subject to national laws. The definition proposed for consideration in the ASEAN region is therefore the "use or threat of violence against civilians, not overtly perpetrated by an official arm of State, to intimidate a public or its institutions in order to achieve a political, religious or ideological objective".

Regional Terrorism Treaty

In 2002-2003, operations were undertaken against terrorists operating in ASEAN countries including Cambodia, Indonesia, Malaysia, the Philippines, Singapore and Thailand. Some of those arrested in Cambodia were Thai nationals, or in Thailand were Malaysian nationals, or in the Philippines and Singapore were Indonesian nationals. Clearly, the ease of movement of people, finance and equipment makes terrorism a cross-border regional challenge. Nor has that challenge yet been met. While certainly making progress in developing counter-terrorism measures, ASEAN has no regional and few bilateral arrangements in place for mutual legal assistance. At the UN Security Council Counter-Terrorism Committee Special Meeting in March 2003, ASEAN acknowledged that there is still much to be done to meet the prescribed benchmarks under Security Council Resolution 1373.

Achievement of those benchmarks relating to legislative reform would be assisted by a well-coordinated regional approach.

A systematically coordinated regional approach has four benefits over ad-hoc mutual legal assistance. Two benefits related to cooperation are: (1) broader extension of mutual legal assistance; and (2) efficiency gains through shared expertise and experience. Two benefits related to harmonisation are: (3) more common transnational mechanisms to support cooperation; and (4) more opportunity to promote best practices in national laws and arrangements. It is obvious that systematic collaboration could be strengthened by a regional framework agreement.

It would facilitate mutual legal assistance and the sharing of resources through an on-going structure. A regional agreement may be the only way to promote rapid harmonisation of approaches to the development of counter-terrorism laws as, in the normal evolution of law, harmonisation accretes very slowly. A regular cycle of meetings would maintain and update cooperation procedures. Following an initial period of law reform to ensure compliance with the common regional approach adopted in the agreement, the Secretariat could identify and circulate information to treaty Parties that would promote the maintenance of best practice through periodic reform.

ASEAN is one of the few geo-political regional groupings not to have adopted a regional counter-terrorism legal framework. Regional treaties have been adopted for Africa, the Americas, Europe, South Asia, the Arab League, the Commonwealth of Independent States and Islamic countries.

Elements for a Regional Treaty

Which of the many possible approaches might ASEAN Member countries take in developing a regional treaty to combat terrorism? Mutual legal assistance in prosecutions? Prevention of terrorist incidents through improved information exchange, financial controls, and migration controls? Should the focus be solely on cooperation to combat transnational terrorism or also address domestic terrorism? Should participation be open to non-ASEAN countries and if so, in what capacity? In essence, these are political questions to be considered in the light of national interests. To an extent, however, they are also legal questions, affected by technical issues of compatibility between national systems of law as well as the mechanics of institutional design.

The definition of terrorist acts set out above provides the conceptual framework for examining whether other regional treaties provide suitable models that might be adapted. A general consideration in considering model provisions for adaptation is that harmonisation of national laws will optimise transnational coordination and cooperation. Specific provisions should also be examined also to ensure that they are clear and fair. The following analysis considers whether the relevant provisions in treaties adopted in other regions meet these criteria.

Formulae for International Cooperation Compared

International agreements to combat terrorism have been formulated at the bilateral, plurilateral and multilateral levels. A study is made here principally of plurilateral agreements, as these are regional and therefore provide the most relevant models for ASEAN Member countries to examine in order to develop their own regional treaty. However, some characteristics of multilateral agreement do inform the purposes of a regional agreement and are discussed as preliminary considerations.

Multilateral Formulae Compared [sh]

There are twelve multilateral treaties adopted under the auspices of the United Nations that pertain to combating terrorism. Three mandate measures to prevent terrorist acts.

Nine were negotiated to respond to a particular type of violent act after the incidents took place. The types of violent acts that are responded to are: attacks on protected persons, hostage taking, hijacking of aircraft, unlawful acts of violence against civil aircraft and airports and offences on aircraft, unlawful acts of violence against ships and against platforms at sea, and bombings of public places.

By narrowly addressing one particular type of violent action, those multilateral treaties are able to criminalise that specific action without needing to define or even refer to terrorism. Therefore they criminalise a specific act but not terrorism generally. Their purpose is simply to deter or punish perpetrators of specific crimes. These multilateral conventions function according to a legal formula to facilitate international cooperation in prosecutions for acts that are, in some way, shared across States.

A keystone in the formula is that there must be an identified transnational factor in the terrorist act that legally engages the jurisdiction of two or more contracting Parties. For example, a hijacked airplane might be registered in Party A, the perpetrator be a national of Party B and the landing take place in the territory of Party C, thus engaging the jurisdiction of three contracting Parties. Conversely, the multilateral treaties do not apply where the act involves no transnational jurisdictional element.

Once the transnational factor has been established, the treaty formula creates rights for, and imposes obligations on, States to establish criminal law jurisdiction over the incident. It requires them either to indict or to extradite the perpetrator. The formula also provides that the treaty may be used as a legal basis for extradition, as might be needed in cases where the Parties have no other extradition agreement between them. Finally, the formula requires States to provide mutual assistance to each other, such as through the exchange of information, to combat the specific category of crime.

In addition to those treaties dealing with particular types of terrorist acts, described above, there are the three first-mentioned multilateral treaties that require preventive action to combat terrorism more broadly. These require national measures that do not entail a transnational factor. Two treaties do not seek to address terrorism directly but to control the distribution of nuclear material and plastic explosives, thus being relevant to the prevention of terrorist acts. The *Convention for the Suppression of the Financing of Terrorism* seeks to prevent the range of terrorist acts listed above. These three multilateral treaties are discussed below in connection with preventive measures together with the regional treaties.

In summary, the multilateral conventions provide two basic models:

1. Obligations to cooperate to extradite or indict persons suspected of having committed specified acts of violence that are transnational and engage the contracting Parties; and

2. Obligations to take national measures to prevent funds and certain dangerous goods from being unlawfully provided to terrorists. Each of these models might have a role in an ASEAN regional treaty.

Regional Formulae Compared [sh]

There are seven plurilateral treaties adopted at regional levels to combat terrorism. In chronological order, they are the:

1. *Organization of American States Convention to Prevent and Punish Acts of Terrorism Taking the Form of Crimes against Persons and Related Extortion that are of International Significance 1971* (OAS Convention);

2. *European Convention on the Suppression of Terrorism 1977* (European Convention);

3. *South Asian Association for Regional Cooperation Regional Convention on Suppression of Terrorism 1987* (SAARC Convention);

4. *Arab Convention on the Suppression of Terrorism 1998;*
5. *Treaty on Cooperation among States Members of the Commonwealth of Independent States in Combating Terrorism 1999* (CIS Treaty);
6. *Convention of the Organization of the Islamic Conference on Combating International Terrorism 1999* (OIC Convention); and
7. *Organization of African Unity Convention on the Prevention and Combating of Terrorism 1999* (OAU Convention).

The 1998 Arab Convention informs the 1999 OIC Convention and the two are virtually identical. Only the OIC Convention is examined here, as it is the later of the two and the parties to the Arab Convention are also parties to the OIC Convention, which has double the membership of the Arab League. The six regional terrorism treaties examined do not follow a single formula to facilitate international cooperation.

Most regional treaties define and set in place a framework for cooperation in the prosecution of terrorists. Unlike the multilateral treaties, they do not necessarily identify and focus on a particular kind of illegal action. Nor do they need a transnational factor in order to trigger international rights and obligations, as most of the multilateral treaties do. They are more diverse in their approaches than the multilateral treaties, indicating the diversity of political perspectives on terrorism amongst the regions in which each was adopted.

The regional terrorism treaties are compared below according to their component functions. The categories of component functions are to: define terrorism (including exceptions); establish national criminal jurisdiction; promote terrorism prevention measures (including intelligence sharing and the exchange of information) as well as mutual assistance in investigations and extradition; and to set in place implementation machinery.

(a) *Definitions of terrorism*: The two main difficulties arising in the regional treaties' various definitions of terrorism are that they cover an inadequate range of violent acts, or that those violent acts defined are not linked to the motivation to influence government. Thus, five of the six treaties do not consistently satisfy the definition of terrorism adopted in this paper that addresses violence against non-combatants and distinguishes common crimes of violence by reason of public motivation. The OAS Convention refers to terrorism but sidesteps any definition. It provides that crimes of kidnapping, murder and assault on protected persons are considered common crimes of international significance, regardless of motive.

Protected persons are domestic and foreign heads of government, senior representatives and diplomats. While the OAS formulation might seem to be a way to sidestep the troublesome question of political motive, it does not meet the wider challenges of regional cooperation to combat terrorist acts against non-combatants because its application is confined to internationally protected persons. The OAS Convention provides that it is up to each individual State to "determine the nature of the acts and decide whether the standards of this Convention are applicable" when determining whether or not to extradite for the commission of a particular offence (Art. 3).

Although it is always a State's task to interpret its obligations (until such time as a dispute is arbitrated), that task is at its most subjective when such vague language renders the obligations uncertain. The OAS Convention leaves it open to Parties not to cooperate simply where they are in sympathy with political motives behind the crime. The European Convention also avoids grappling with the question of political motive. It does this by listing offences that shall not be regarded as political offences for the purposes of granting asylum (Art. 1). The list is confined to violent acts, some of which are already the subject of cooperation under multilateral instruments, and some that are more general.

Acts covered under the United Nations treaties on unlawful acts of violence against civil aircraft and on aircraft hijacking, physical attacks on internationally protected persons and hostage taking are listed. Beyond the multilateral treaties already in place in 1977, the year of adoption of the European Convention, bombing and explosives offences are listed. The European Convention has the limited but useful feature of ensuring that the listed multilateral treaties apply to all regional Parties. However, because the multilateral treaties cross-referred to are set out in the body of the text of the European Convention, rather than in annexes for which simplified amendment procedures might be adopted, it has not kept up to date with the multilateral treaties.

For example, the bombing and explosives offences listed, although extending beyond the scope of multilateral treaty definitions of terrorist offences as set out in 1977, have been overtaken by the more extensive provisions of the *International Convention for the Suppression of Terrorist Bombings* in 1997. This anachronism demonstrates the shortcomings of a definition based on other extant treaties that do not themselves define terrorism in general terms. Article I of the SAARC Convention provides a list of conduct which shall be regarded as "terroristic" and which shall not be regarded as a political offence for the purposes of granting asylum.

These include offences under listed multilateral treaties. This part of the provision generalises the application of the listed multilateral treaties throughout the SAARC Member States. In addition, any treaty to which particular SAARC Member States concerned are already Parties and that obliges them to prosecute or grant extradition is covered. Other offences which are deemed to be "terroristic" include murder, manslaughter, assault causing bodily harm, kidnapping, hostage-taking and offences relating to firearms, weapons, explosives and dangerous substances when used to perpetrate *indiscriminate* violence involving death or serious bodily damage to persons or serious damage to property.

Conspiracies to commit the above are also included within the scope of the Convention. This latter part of the provision avoids the question of motive by addressing ordinary violent crime rather than terrorism. The qualifying descriptor, requiring that the violence be *indiscriminate*, creates an unsatisfactory ambiguity, allowing for a subjective interpretation that political violence is not indiscriminate. Thus, this aspect of the legal definition for terrorism is unclear and does not address the need for a public motivation. The CIS Treaty defines terrorism in Article 1 as an illegal act punishable under criminal law, committed for the purposes of undermining public safety, influencing decision-making by authorities, or terrorising the population.

The kinds of illegal acts listed as the threat of or actual violence are: violence against natural or juridical persons; destruction or damage to property so as to endanger peoples' lives; causing substantial harm to property; action against the life of a statesman or public figure for the purpose of putting an end to either his State or a public activity or for revenge for such activity; and attack against representatives, premises or vehicles of a foreign State or international organisation. Other acts classified as terrorism are those recognised under universally recognised international legal instruments on counter-terrorism. Oddly, any acts defined as terrorism under the national legislation of CIS Parties are included by reference. This introduces an unknowable factor in the treaty definition and creates an uncertainty that is inherently unsatisfactory. Article 1 also defines the term "technical terrorism".

This refers to the use or threat to use nuclear, radio-logical, chemical or biological weapons, if the acts are for the purpose of undermining public safety, terrorising the population or influencing the decisions of the authorities in order to achieve political, mercenary or other ends. Although aspects of the CIS definition, as set out in Article 1 conform to the definition proposed in this paper, it introduces limitations and wide uncertainties inconsistent with the proposed definition, such as by its referential inclusion of multilateral conventions and national legislation. The definition's structure is confusing and repetitive. The definition of "technical terrorism" does not seem to add any new conceptual or legal elements.

Article 1(2) of the OIC Convention defines "terrorism" as any act or threat of violence, notwithstanding its motives, perpetrated to carry out a criminal plan to terrorise people, threaten to harm them or endanger their lives. It goes on to extend the elements of such a plan to include also harming people's honour, freedoms, security or rights, exposing the environment or any public or private property to hazards, occupying or seizing property, endangering national resources or international facilities, or threatening the stability, territorial integrity, political unity or sovereignty of any State. These latter acts are extensive but inadequately defined. For example, the notions of harm to people's rights or of exposure of the environment to hazards are extraordinarily vague and do not adequately describe acts of violence.

The inclusion of threats to the political unity of a State would encompass separatist movements regardless of their peaceful means, pre-judging the legitimacy of political motivation. Thus, in addition to being vague, the OIC definition has a steep political slant and is not objective. Article 1 goes on to distinguish between "terrorism" and "terrorist crimes". Article 1(4) lists crimes stipulated in various multilateral conventions and deems those acts to be "terrorist crimes", but only if the conventions were ratified by the Party concerned. The utility of the cross-reference to multilateral conventions is undercut by excluding application to non-Parties to those conventions. Article 1(3) of the OIC Convention defines "terrorist crimes" more broadly as any crime perpetrated, commenced or participated in to realise a terrorist objective in any Contracting State or against its nationals, assets, interests or foreign facilities, punishable by its national laws.

The prerequisite of criminality under the national legislation of the OIC Parties renders the definition subjective. Article 2(d) stipulates that all "international crimes" aimed at financing terrorist objectives shall themselves be classed as terrorist crimes. This part of the definition extends the meaning of "terrorist crimes" to include other criminal activities, such as trafficking in narcotics and human beings, that are intended to finance terrorist plans. The meaning of "international crimes" is not clear and does not address the financing by simple donation of funds by supporters, perhaps the most ubiquitous form of financing.

It casts "'terrorism", "terrorist crimes", "international crimes", as well as ordinary and political crimes, together into one conceptual swamp. The vague sweep of this definition is unsatisfactory. The most recent of the regional treaties is the OAU Convention. Article 1(3) of the OAU Convention defines a "terrorist act" as any act which violates the criminal law of a State Party and which may endanger the life, physical integrity or freedom of, or cause serious injury or death to any person or a group of people. It includes acts that cause damage to public or private property, natural resources, and environmental or cultural heritage.

The acts must be intended either to intimidate any government into doing or abstaining from doing something, disrupt any public service, or to create general insurrection in a State. Thus, consistent with the proposed definition, the violence is distinguished from common crime by its political motive. Promoting, sponsoring, organising or in any other way contributing to a terrorist act, as defined in Article 1(3)(a), is itself part of a terrorist act (Art. 1(3)(b)). The confined range of preliminary acts is too narrow to pick up financing of terrorist acts prior to a plan being finalised or of terrorist organisations as such. However, with the exception of its precondition of violation of national criminal laws, it is similar to that proposed in Section 1 of this paper and is the clearest and most usefully applicable definition adopted among the regional treaties.

Overall, the language used in the regional treaties to define terrorism is often ambiguous, qualified and unsatisfactory, leaving much room for non-cooperation. A common theme is the limited coverage of violent acts in the list of prohibited acts. Often the list of acts is extremely limited and constrains the opportunities for international cooperation. In several instances, however, the acts listed are broad and vague, covering ordinary transnational crime without regard to the contextual elements of armed conflict, or else, covering political opposition irrespective of violence against civilians intended to induce a state of fear. There is a chronological trend away from limited listed acts and towards general principles to create a definition.

The latest treaty, the OAU Convention, provides a useable definition, the main weakness of which is that the particular criminal act must already be defined as a crime under national law. The circularity of this prerequisite is discussed below.

a. Finally, the *Convention for the Suppression of the Financing of Terrorism*, exceptionally among the multilateral conventions, employs a general definition of terrorism, in order to supplement the violent acts defined in other multilateral conventions to which it cross-refers.

Article 2.1(b) defines an offence within the scope of the Convention as being: Any other act intended to cause death or serious bodily injury to a civilian, or to any other person not taking an active part in the hostilities in a situation of armed conflict, when the purpose of such an act, by its nature or context, is to intimidate a population, or to compel a government or an international organization to do or to abstain from doing any act. The Financing of Terrorism Convention provides a model definition that sets in place a robust conceptual framework.

The congruity and minor departures and elaborations are: (1) it defines serious violence as requisite but limits the acts of violence to those intended to cause death or bodily injury, thereby excluding attacks on public infrastructure and threats; (2) in addition to civilians, the violence is directed at non-combatants; (3) the intention to coerce a social institution in the form of government is required, but also specified is coercion of an international organisation or intimidation of a population; and (4) the character of the perpetrator is identified as that of an individual person, *i.e.*, a non-State actor.

The Convention has been ratified by all ASEAN Member States, except Malaysia and Laos. Its definition is, therefore, likely to be acceptable for adaptation and adoption into an ASEAN regional treaty.

b. *Exceptions for certain political motives*: The definitions of terrorism formulated in each of the six regional treaties avoid evaluation of political motive. They specify that the offences described shall not be considered political offences, thereby excluding the possibility of offenders being granted political asylum.

On the other hand, some of the treaties exclude from the scope of their definition violence for certain political purposes. Thus, Article 2(a) of the OIC Convention provides that a peoples' struggle including armed struggle against foreign occupation, aggression colonialism, and hegemony, aimed at liberation and self-determination in accordance with principles of international law shall not be considered a terrorist crime. This exception contemplates that acts that would otherwise be considered terrorist are condoned for a wide range of purposes. The purposes are broad and imprecise.

Although crafted primarily to condone acts of terror against Israeli interests, in the new millennium, the exception could also work against the interests of those OIC States, such as Algeria, Egypt, Indonesia, Morocco, Saudi Arabia and Uganda, that are threatened by rebels that engage in terrorist acts. Article 3(a) of the OAU Convention similarly states that "the struggle waged by peoples in accordance with the principles of international law for their liberation or self-determination, including armed struggle against colonialism, occupation, aggression and domination by foreign forces shall not considered as terrorist acts" (sic).

For those terrorist acts not excepted, it provides that there is no "justifiable defence" for political, philosophical, ideological, racial, ethnic or religious motives (Art. 3(b)). The difference between violent actions engaged in for self-determination, as compared to acts for political or ideological motives, is too obscure to be applied other than subjectively. Clearly, these exceptions allow terrorist acts for certain political and ideological purposes because they fail to distinguish them from legitimately conducted armed conflict under international law.

The exceptions are ambiguous and open to subjective interpretations that could arbitrarily exclude almost all violent political conflicts. In the ASEAN region, some might consider that they provide exceptions for organisations including the Moro Islamic Liberation Front, Abu Sayyaf Group, New Peoples Army, Pattani United Liberation Organization, Kumpulan Mujahidin Malaysia, Gerakan Aceh Merdeka, Laskar Jihad, Islamic Jihad, *etc.* Thus, the exceptions formulated give ample scope for Parties to avoid cooperation under a treaty and are unacceptable model provisions.

c. *Criminalisation under national jurisdiction*: The multilateral terrorism treaties require an identified transnational factor in a terrorist act that connects and legally establishes a State's jurisdiction, creating rights and obligations for the State concerning the incident.

For example, a Party might be required to establish procedures and penalties in its criminal law for the defined act and to indict or extradite the perpetrator. In contrast, in regional terrorism treaties a variety of conditions may predicate criminal jurisdiction. Some require a transnational factor, particularly when the regional treaty's definition of a terrorist act cross refers to multilateral treaties. Other regional treaties may create obligations to establish jurisdiction for terrorist acts that are entirely domestic but, in deference to sovereign independence, premise their definitions of terrorist acts on the breach of a previously established national criminal law. Although most regional treaties impose an obligation to establish new prohibitions, procedures and penalties at criminal law, they all defer to national sensitivities by creating only soft, obtuse obligations.

Under Article 8(d) of the OAS Convention, the Parties are to cooperate in endeavouring to ensure that the listed acts are included in each Party's penal code. European Convention Parties are cryptically required to establish criminal jurisdiction over the offences listed in Article 1 where the offender is present in its territory (Art. 6.1). Parties perform their obligations under the SAARC Convention only to the extent permitted by their national laws, as is apparent in Arts. V, VI and VIII. Article 6 of the CIS Treaty requires the Parties, through consultations, jointly to draw up mere recommendations for "achieving concerted approaches" to the legal regulation of issues relating to combating terrorism.

The shortcomings in these obligations are manifest. The OAS commitment is a "soft" or voluntary obligation, and the European requirement is to assert procedural jurisdiction rather than to criminalise an act. Under the SAARC and CIS treaties, there is no obligation at all to adopt laws criminalising terrorist acts. Indeed, Article 9(2) of the CIS Treaty states that mutual assistance may be denied where the act in relation to which the request was made is not a crime under the legislation of the requested Party. Under the OIC Convention, Article 3(II) provides that Parties are committed to preventing and combating terrorist crimes in accordance with the provisions of the Convention and their respective national rules.

But, as noted above, "terrorist crimes" in the OIC Convention is defined under Article 1(3) as any crime perpetrated, commenced or participated in to realise a terrorist objective *already punishable by national laws*. Due to this circular formulation, there is no obligation in the Convention on Parties to adopt applicable national laws. Parties to the OAU Convention undertake in Article 2 to review their national laws and establish criminal offences for terrorist acts as defined by the Convention. Part III of the Convention outlines State jurisdiction over terrorist acts and provides that, upon receiving information that a person who has committed a terrorist act may be present in its territory or is one of its nationals, then the Party must take measures under its national law to investigate and prosecute the person. Parties have the option of establishing jurisdiction if the terrorist act is committed against a national, stateless resident, property or against the security of the State (Arts. 7(1) and 7(2)).

It is apparent that the OAU Convention affords the broadest agreed base for establishing national laws and that it imposes the most significant rights and obligations in relation to criminal law enforcement. However, although there is an apparent requirement to establish criminal offences for terrorist acts, a particular action only falls within the definition of terrorist acts if it is already a violation of the criminal laws of the Party (Art. 1(3)(a)). Thus, the requirement to establish criminal jurisdiction is circular.

Consequently, there is no clear requirement to enact legislation to cover actions that are peculiar to terrorism and not normally covered by ordinary criminal law. Such actions not normally covered might include, for example, bombings, extraterritorial violence, surreptitious funding, membership in terrorist organisations, espionage, and threats and hoaxes.

An ASEAN regional treaty could build on the particular strengths of the OAU Convention provisions, while avoiding its circularity. Its confusion concerning the criminalisation of terrorist acts under national law works against the adoption of a common definition and the harmonisation of national laws. This could be resolved by formulating obligations that require treaty Parties to enact legislation to criminalise terrorist acts as set out in the treaty definition. That definition should not include a prerequisite that an action first be a violation of criminal law in order for it then to be a "terrorist act".

A template for the offences to be criminalised could include appropriate preliminary and accessory acts. Clear, specific obligations to adopt mandatory procedures to establish criminal jurisdiction are then needed to indict or extradite perpetrators.

d. *Prevention measures*: Measures to prevent terrorist acts can entail a wide range of governmental activities, including inter-agency coordination, immigration controls, customs controls, financial flow controls and the securing of public places. The regional treaties address both domestic and international prevention measures, with emphasis on international cooperation. The range of areas for prevention addressed varies widely across the treaties, demonstrating a roughly chronological progression towards more elaborate prevention measures, with the OAU Convention having the most extensive coverage. Nevertheless, most treaty measures are couched in the language of soft obligations and, often, in vague terms.

The OAS Convention provides that the Parties are to "cooperate among themselves" to "prevent and punish acts of terrorism, especially kidnapping, murder, and other assaults..." (Art. 1). Article 8(a) loosely provides that the Parties accept the obligation to take measures to prevent and impede the preparation in their territories of the crimes mentioned in Article 2. The European and SAARC treaties impose no dedicated prevention obligations. Part II of the OIC Convention sets out extensive areas for cooperation to combat and prevent terrorist crimes. Article 3.I calls upon the Parties not to support terrorist acts and Article 3.II (A) lists the preventative measures that each Party "shall see to".

These include barring their territories from being used as arenas for planning, organising and executing terrorist crimes; developing and strengthening border control and surveillance on transfer or stockpiling of weapons; and strengthening security of protected persons, international organisations, vital installations and public transport facilities. Part II of the OAU Convention also specifies areas of counter-terrorism cooperation. Parties undertake to refrain from acting in such a way as to support the commission of terrorist acts, including the issuing of visas and travel documents (Art. 4(1)).

All Parties must adopt "legitimate measures" to prevent and combat terrorism in accordance with their national structure, including the development and strengthening of monitoring for activities such as arms stockpiling, strengthening the protection and security of diplomatic and consular missions and international persons and promoting the exchange of information and expertise on terrorist acts (Art. 4(2)). Article 2 of the CIS Treaty provides that the Parties shall cooperate in preventing, uncovering, halting and investigating acts of terrorism, in accordance with the Treaty, their national legislation and any international obligations.

The inclusion of national legislation seems to qualify the provision and render it indeterminate. Article 5(1) (c) briefly elaborates that the Parties shall assist one another by developing and adopting agreed measures for preventing, uncovering, halting or investigating acts of terrorism, and informing one another about such measures and Article 5(1)(d) goes on to provide that they shall adopt measures to prevent and halt preparations in their territory for the commission of acts of terrorism in the territory of another Party. The OIC and OAU Conventions usefully identify common areas for cooperation in preventive measures, being the stockpiling of weapons, border controls over dangerous goods and suspect persons, and security for protected persons.

None of the regional treaties address the prevention of terrorist financing and none provide strong models for the implementation of preventive measures. An original approach to border controls would be to insert a commitment between Parties to an ASEAN treaty to develop a protocol for cooperation in border measures that requires the development of immigration, customs and financial controls. These could include the sharing of alert lists and the designation of national coordination and communication points to facilitate international cooperation in preventive measures.

In relation to protected persons, the treaty might require that the Parties hold periodic consultations to identify security measures needed for particular protected persons. Joint crisis management plans could also be required to be formulated. In connection with the stockpiling of weapons and the financing of terrorists, three multilateral conventions concern prevention measures that provide useful models. The *Convention on the Marking of Plastic Explosives for the Purpose of Detection* obliges each Party "to take the necessary and effective measures to prohibit and prevent the manufacture in its territory of unmarked explosives" (Art. II) as well as "the movement into or out of its territory of unmarked explosives" (Art. III).81 Explosives must be marked by adding a detection agent to them during the manufacturing process (Art. I.3).

The terms "explosives" and "detection agent" are defined in the Technical Annex to the Convention (Arts. I(1) and I(2)). Article IV obliges the contracting Parties to destroy unmarked explosives in their territory other than those explosives in the possession of the police or the military (Art. IV(4)). The Plastic Explosives Convention is the multilateral convention least ratified by ASEAN Member States. All six regional conventions contain provisions relating to cooperation between contracting Parties, which vary in detail. An ASEAN regional treaty might elaborate on the regional conventions and provide a more comprehensive system of cooperation and education. Articles in four regional conventions contain relevant provisions that extend beyond the obligations in the Plastic Explosives Convention, concerning the stockpiling of explosives, weapons and ammunition.

Only the OIC Convention specifically states that a contravention of the Plastic Explosives Convention is a terrorist crime, provided that the contracting State is a party to that multilateral convention (Art. 1.4(l)). The *Convention on the Physical Protection of Nuclear Materials* obliges States to control access to nuclear material and equipment. Article 3 imposes the general obligation on Parties to take steps to ensure that, during international transportation, nuclear materials (defined in Annex II) within their territory or on board a registered ship or aircraft travelling to or from that State are protected. The effect of the Convention is that Parties may not import from non-contracting Parties, or export to Parties or non-contracting Parties, unless the nuclear material is protected according to the standards set out in Annex I (Art. 4). Article 5 specifies how Parties are to cooperate in the event that nuclear material is unlawfully taken.

In general, they are obliged to criminalise within their national laws the unlawful taking of, or any threat to use nuclear material to kill or cause serious injury or damage (Art. 7), to establish jurisdiction over the offences (Art. 8), and to prosecute or extradite offenders (Art. 10). The Parties are to cooperate concerning criminal proceedings brought in respect of the offences (Art. 13.1). The only regional treaty to deal specifically with nuclear technology, although to little effect, is the CIS Treaty. The OIC Convention also provides that offences under the Nuclear Materials Convention are terrorist acts, provided that the latter instrument has been ratified by the relevant Parties.

Although the regional conventions embrace, in general terms, the definition of an offence as set out in the Nuclear Materials Convention, they do not cover its preventive measures for the protection of nuclear materials. Therefore, there would be advantage in importing the Nuclear Materials Convention obligations into an ASEAN regional treaty by cross-references to it. The broadest in application of the multilateral conventions on prevention is the *Convention for the Suppression of the Financing of Terrorism*.

It aims to deprive terrorists of their sources of "funds", which are any type of asset, regardless of how acquired (Art. 1.1). A person commits an offence within the meaning of the Convention if he or she "by any means, directly or indirectly, unlawfully and wilfully, provides or collects funds with the intention that they should be used or in the knowledge that they are to be used, in full or in part, in order to carry out" the defined terrorist offences (Art. 2.1). Regional provisions to suppress the financing of terrorism are found in the OAU Convention, where Parties undertake to refrain from financing terrorist acts (Art. 4.1). A terrorist act is defined to include "sponsoring" (Art. 3(b)).

The OIC Convention also contains relevant provisions but merges money laundering with terrorist acts where it states that money laundering aimed at financing terrorism is itself to be considered a terrorist crime (Art. 2(d)). It would seem a better approach to prosecute perpetrators of money laundering activities intended to finance terrorist acts under the two separate charges of money laundering and of terrorist financing so as to maintain the conceptual distinctions between money laundering and terrorist financing. The OIC Parties also agree that their State organs will not "execute, initiate or participate in any form in organizing or financing or committing or instigating or supporting terrorist acts whether directly or indirectly" (Art. 3(1)).

However, the OIC Convention does not address private fund raising for terrorist acts and it would seem that the Financing of Terrorism Convention, both in its definition of "funds" and in proscribing the offence of "financing", is broad enough to encompass even small donations by individuals where they indirectly lead to the commission of a listed offence (Art. 2.1), provided that the person intended or knew that such money would be so used. Further, an individual's donation need not be actually used in the commission of a terrorist offence if such money was contributed to the organisation's pool of assets with the intention that it be used for the purposes of terrorism (Art. 2.3).

Presuming that Malaysia and Lao will ratify the Financing of Terrorism Convention, its provisions are sufficient to serve the purpose of regional suppression of terrorist financing. Although ASEAN Member States could usefully apply as between them the Plastic Explosives and Nuclear Materials Conventions, incorporating the obligations by reference into a regional treaty annex, these should be supplemented to develop and apply border controls on the flows of weapons and dangerous goods. Clearer obligations and procedures in relation to the extradition of offenders and capacity building are discussed.

e. *Intelligence information exchange*: Intelligence is taken here to mean information gathered about terrorist acts in advance of their commission.

The gathering and exchange of intelligence is a part of the set of preventive measures needed to combat terrorism but is considered separately because of its central importance in international cooperative efforts. The sharing of intelligence is explicitly provided for in most regional terrorism treaties. However, the agencies, methods and protocols used in intelligence gathering and analysis are not specified, consequent upon the sensitivity of this governmental activity. Under Article 8(b) of the OAS Convention, Parties agree to cooperate in preventing and punishing the listed crimes, including through the exchange of information. SAARC Parties are obliged to cooperate amongst themselves in relation to the exchange of information with a view to preventing terrorist activities (Art. VIII(2)).

The OIC Convention Parties must cooperate amongst themselves in exchange of information, as well as investigation, exchange of expertise and education (Arts. 3.II(A)7-8 & Art. 4). Under Article 5 of the OAU Convention, Parties undertake to strengthen the exchange of information amongst themselves regarding acts and crimes committed by terrorist groups and the communication methods used by such groups (Art. 5(1)). Both the OIC and OAU Conventions oblige the Parties to respect the confidentiality of any information passed to them. These minimalist obligations stand in contrast to those of the CIS Treaty. Article 5(1)(a) of the CIS Treaty provides that Parties shall cooperate and assist one another by exchanging information. Article 5(1)(h) provides for the exchanging of legislative texts and materials.

Article 11 provides that the Parties shall exchange information on issues of mutual interest. These include: materials distributed in their territories containing information on terrorist threats; acts of terrorism in the course of preparation; illegal circulation of nuclear, chemical or biological weapons and the like; terrorist organisations or individuals that present a threat to the national security of one of the Parties; illegal armed formations employing terrorist methods; ways, means and methods of terrorist action they have identified; supplies and equipment that may be provided by one Party to another Party; practice with respect to the legal issues that are the subject of the Convention; identified and presumed channels of terrorist financing and suppliers; and terrorist encroachments aimed at violating the sovereignty and terrorist integrity of Parties.

The relatively high degree of specificity might be attributable to the historic character of the CIS as the USSR, being one polity under the influence of Russian security institutions, and the common nature of their contemporary threat environment. The diversity of conditions — political, cultural, economic, and religious — between ASEAN members suggests that intelligence sharing will be approached cautiously and that highly specific and mandatory provisions would not be appropriate. Ultimately, States exercise full control over the intelligence that they gather and will choose to share it at their discretion.

As the information can be classified at various levels of secrecy, State implementation of treaty obligations will not be transparent. Nevertheless, explicit obligations to set up international contact points for "24/7" coordination of information and to treat that shared information as confidential are, at least, useful to flag the conditions that encourage intelligence sharing.

f. *Mutual assistance for investigations*: Investigations concern information gathering (*e.g.*, documents, communications intercepts, exhibits) and rendering of persons (*e.g.*, witnesses and suspects) for the purpose of law enforcement in response to criminal acts already committed.

In relation to investigations, the focus of the regional treaties is on mutual assistance. Their relevant provisions are of a highly general nature, except in relation to extradition, which is dealt with in more detail below. The

OAS Convention makes no provision at all for investigations. Article 8 of the European Convention obliges States to "afford one another the widest measure of mutual assistance in criminal matters in connection with proceedings brought" in respect of terrorist offences. SAARC Convention Parties are required to afford one another mutual assistance in connection with proceedings brought in respect of terrorist offences, including supplying evidence where necessary (Art. VIII(2)).

Under Article 5(1)(b) of the CIS Treaty, Parties shall cooperate by responding to inquiries on the conduct of investigations. Article 9(1) provides that the rendering of assistance shall be wholly or partially denied if the requested Party believes that cooperation may impair its sovereignty, security, social order or other vital interests or may contravene its legislation or international obligations. Article 3.II(B) of the OIC Convention lists "combating measures" as an area for cooperation that includes arresting perpetrators of terrorist crimes, ensuring protection of witnesses and investigators to terrorist crimes, and establishing effective cooperation between the concerned government agencies and the citizens for combating terrorism.

Article 4 requires Parties to promote cooperation with each other in the field of investigation procedures, and particularly in relation to arresting escaped suspects or those convicted of terrorist crimes. Article 14 provides that each Party shall extend to the others every possible assistance for investigation or trial proceedings related to terrorist crimes. OAU Convention Parties undertake to exchange information leading either to the arrest of any person charged with a terrorist act against the interests of a Party, or to the seizure of arms (Art. 5(2)). Part V of the Convention outlines procedures in relation to extra-territorial investigations and mutual legal assistance. Article 14(1) provides that any Party may request another Party to carry

(2005) out, with its assistance and cooperation, on the latter's territory, criminal investigations related to judicial proceedings concerning alleged terrorist acts.

Apart from some minor mentions in the OIC and OAU Conventions, the regional treaties do not specifically address procedures for inter-jurisdictional taking of evidence, transfer of foreign persons to give evidence or to assist in investigations, service of judicial documents, or execution of search and seizure. Although some of these matters may already be addressed adequately under other international arrangements for cooperation in combating crime, there is no cross-reference in the regional treaties to such arrangements. A regional terrorism treaty is an inappropriately narrow base for a framework for mutual assistance that would be required in a wide range of other criminal investigations. Mutual assistance procedures tend to be detailed and technical, too much to be incorporated into the text of a framework treaty narrowly addressing regional terrorism.

Therefore, broad ranging, detailed technical arrangements for mutual assistance are better developed outside regional terrorism treaties. However, where no regional mutual assistance arrangements have been established, the regional terrorism treaty could impose an obligation on Parties to develop them. At the very least, Parties should undertake to review their mutual assistance arrangements to ensure that they are coordinated with and complement counter-terrorism treaty obligations. No regional mutual assistance legal framework currently exists for the ASEAN region. A Malaysian proposal to develop a framework is currently in the early stages of development.

Ideally, an ASEAN regional terrorism treaty should be coordinated within a broader ASEAN regional mutual assistance framework. In the event that ASEAN countries do not adopt a regional framework for mutual assistance, the terrorism treaty should require that they develop bilateral arrangements for mutual assistance to investigate terrorist acts.

g. *Extradition*: The historic focus of multilateral legal cooperation to combat terrorism has been on the obligation to indict perpetrators or to extradite them to jurisdictions that will do so. The obligation to indict, as set out in regional treaties, was discussed above under the heading of "Criminalisation Under National Jurisdiction". Legal issues that arise in relation to extradition concern whether to classify an offence as essentially political, in which case obligations to extradite do not apply.

If the request is considered to be persecution for political activity, the State with custody over the accused has the right not to extradite, and may instead be obliged to grant asylum. The OAS Convention provides that persons charged or convicted with the listed crimes shall be subject to extradition under the provisions of extradition treaties in force between the Parties (Art. 3). The Parties undertake to include these crimes among the punishable acts giving rise to extradition in any treaty to which the Parties may agree in the future (Art. 7). Where extradition is not possible, Parties are obliged to submit the offender to its competent authorities for prosecution (Art. 5).

These provisions are qualified in that the OAS Convention provides that it is not to be interpreted so as to impair the right to asylum (Art. 6). Because there is no general definition for terrorist offences in the OAS Convention, this asylum qualification is wide open to subjective interpretation that the alleged offences are merely political. It undermines the extradition obligation and much of the Convention. The European Convention Parties are required either to extradite persons accused or to prosecute them (Art. 7). Extradition may be refused on the grounds of granting political asylum (Art. 5). This exception is also drafted in broad terms that render it indeterminate and open to abuse.

Further to the listed terrorist offences, Parties are free to decide that a 'serious act of violence against the life, physical integrity or liberty of a person' should not be classed as a political offence for the purposes of extradition (Art. 2). That decision is, of course, a choice that they would have independent of the Convention. SAARC Convention Parties are required to extradite, subject to their national laws (Art. VI). This subjection to national laws undermines the obligation, which is, anyway, subject to a range of further broad qualifications for triviality, inexpediency, injustice and bad faith (Art. VI).

Unusually, the CIS Treaty makes no independent provision for extradition. Article 5(2) simply states that extradition procedures shall be determined by the international agreements to which the States concerned are Parties. Article 4(2) provides that the nationality of the person accused for an act of terrorism shall be deemed to be his nationality at the time of the commission of the act. Under the OIC Convention, Parties undertake to extradite those indicted or convicted of terrorist crimes (Art. 5). Extradition may be refused if the crime for which extradition is requested is deemed, under the laws in the requested State, to be one of a political nature (Art. 6.1).

However, Article 2(b) declares that terrorist crimes shall not be classed as political crimes. Despite this broad basis for extradition, it should be remembered that the OIC Convention exception from its definition of terrorism is wide open. Article 23 provides that any request for extradition must be accompanied by the original or an authenticated copy of the indictment or arrest order issued in accordance with the conditions stipulated in the requesting State's legislation. Parties must also provide a statement of the acts for which extradition is sought, which specifies details such as dates and places, and a description of the subject wanted for extradition.

These procedural requirements are specific, which is advantageous in the absence of other extradition arrangements. Part IV of the OAU Convention deals with extradition. Article 8(1) provides that Parties shall extradite a person charged with or convicted of a terrorist act defined in Article 1 and carried out within the territory of another Party. However, the definition of terrorism is again subject to broad exception. The extradition must be requested by one of the Parties in conformity with the rules outlined in the Convention, which set out procedural matters and do not allow for subjective exceptions (Arts. 9-13). It is apparent that the OAU Convention provides the most clear and objective standards for extradition. Distinguishing legitimate political acts and armed struggle from terrorist acts is the path through the thicket of confusion between extradition and asylum obligations.

Political violence that meets the criteria in the definition of terrorism is neither legitimate political activity nor legitimate armed conflict. Thus, once a regional treaty has adopted a clear definition for terrorist acts, that definition provides the path for extradition obligations. Legitimate exceptions based on asylum to the obligation to comply with requests for extradition of alleged terrorists should be permitted only in cases of *mala fides* on the part of the requesting State, as determined by an international tribunal should one have jurisdiction. The regional treaties offer several approaches to building a procedural basis for extradition.

The OAS Convention and CIS Treaty cross-refer to extradition treaties between their Parties, whereas the OIC and OAU Conventions introduce their own procedural formalities. The European and SAARC Conventions are silent as to procedure. It would seem appropriate to cross-refer to procedures in extradition treaties already in place, as these will set out processes specifically tailored to serve the parties' particular needs. However, as extradition treaties are mostly bilateral, it can be expected that there will gaps in regional coverage. A regional terrorism treaty might, in fact, be the only extradition instrument in place between some countries and it should therefore also provide simple default procedural formalities for extradition processes.

h. *Measures to improve compliance*: Efforts to gather and analyse relevant intelligence, to take effective preventative measures, and to investigate successfully and to prosecute criminal acts require sophisticated prevention and enforcement capacities.

Around the world these capacities are under development, often in the early stages. Under the European Convention, the Council of Europe's Committee on Crime Problems is to be kept informed on the implementation of the Convention (Art. 9(1)). SAARC Parties are obliged to exchange intelligence and expertise (Art. VIII(2)). The OIC Convention provides for the exchange of expertise and for Parties to cooperate with each other to undertake and exchange studies and research on combating terrorist crimes. They are also to provide each other with technical assistance within the scope of their capabilities (Art. 4).

The OAU Convention calls for cooperation in relation to the provision of technical assistance between the Parties (Art. 5(6)). These provisions compare poorly with those on regional capacity building found in the CIS Treaty. Article 5(1)(f) makes provision for the joint financing and conduct of research and development work on systems and facilities posing technological and environmental danger. Articles 5(1)(g) and 12 allow for special anti-terrorist units to give practical assistance in preventing terrorism and dealing with its consequences, by agreement between interested Parties.

Article 5(1)(h) allows for exchanging experience on the prevention and combating of terrorist acts through training courses, seminars and workshops. Article 5(1) (i) calls for the cooperation of the Parties through training and further specialised training of personnel. Article 7 provides that cooperation under the CIS Treaty shall be conducted on the basis of requests by an interested Party for assistance to be rendered, or on the initiative of a Party that believes that such assistance would be of interest to another Party. Article 8(1) provides that the requested Party shall take all necessary measures to ensure the prompt and fullest possible fulfillment of the request.

These latter provisions are the most specific and mandatory found in the regional treaties. In practice, of course, their implementation relies on mutual good will and the measures outlined serve primarily as expressions of good will and encouragement. International agreements can employ more sophisticated measures to facilitate capacity development by setting reporting obligations and performance benchmarks for Parties. While exchanges of expertise are encouraged, or expressed in almost mandatory terms in the CIS Treaty, only the European Convention contains obligations to report on implementation.

None of the treaties establish institutions or refer to coordination with other institutions to promote capacity building in the field of counter-terrorism. For example, conferences of Parties or their relevant sub-committees could be mandated with responsibilities to receive and review prescribed reports on treaty implementation. They could recommend development or amendment of procedures for enhancing national implementation and international cooperation. In particular, they could be mandated to develop, adopt and oversee the implementation of detailed action plans for technical capacity development, such as legislative and administrative reforms, so as to give political impetus to a coordinated regional program of prevention and enforcement capacity building.

Therefore, the initiation of institutional mechanisms to provide oversight of regional capacity building should be a feature of an ASEAN regional treaty. Another issue related to the integrity of the implementation of a treaty is the design of its procedures for dispute resolution, if any. These may prove necessary to resolve questions of implementation, for instance, where an extradition request is refused. Acute political sensitivity will pervade such refusals, as terrorist acts seek to undermine the public institutions in the victim State. Accordingly, Parties to regional treaties have not chosen to subject disputes between them to formal and binding dispute resolution procedures.

ASEAN Member States are also most unlikely to opt for formal and binding dispute resolution. Finally, it should be noted that two of the most recent regional treaties do not allow Parties to make reservations to their obligations. This is a desirable trend for ASEAN Members to adopt as it protects the function of the agreement.

Outstanding issues

There are many matters pertinent to efforts to combat terrorism that are not covered in the above examination, which is limited to matters arising within the four corners of the regional treaties. But two questions that would confront future drafters of an ASEAN regional treaty that remain outstanding are briefly noted here. They concern safeguards for civil and political rights and legal relationships with non-Parties. The safeguards issue is raised in the OAS Convention, which gives any person deprived of his or her freedom through application of the Convention the right to enjoy legal guarantees of due process (Art. 4). Article 8(c) gives every offender the right to defend himself or herself.

The OIC Convention outlines measures for protecting witnesses and experts, although its design may work against the administration of justice as penalties are not to be inflicted upon witnesses or experts who do not comply with a summons. These two treaties demonstrate no coherent approach to the safeguarding of civil rights. Better safeguards that could be considered include a saving, in the intention component of the definition of terrorist act, to protect the innocent organisers of *bona fide* political demonstrations that nevertheless turn violent. Appropriate guarantees of humane treatment, due process and legal representation for suspects could also be included. Another issue related to the fair administration of justice is that of sentencing.

Regionally accepted national sentencing guidelines may prove useful in facilitating cooperation where a Party is reluctant to extradite because it considers applicable penalties in the requesting State to be too harsh (or too light). Sentencing is not addressed in any of the regional terrorism treaties, although the seizure of terrorist assets is dealt with in the OIC and OAU Conventions. The second question, concerning legal relationships with non-Parties, has several aspects. These include whether an ASEAN treaty should oblige its Parties to observe the terms of the multilateral counter-terrorism treaties and whether an ASEAN treaty should be open to participation by non-ASEAN States.

It is possible, by cross-reference, to import into an ASEAN regional treaty the commitments set out in all the multilateral treaties. Of the few regional treaties that import obligations, only one imposes upon all its Parties the obligations set out in multilateral treaties. The device of importing obligations has been sub-optimal in practice, however, as most regional treaties lack efficient amendment procedures essential to keep them current with multilateral developments. Where obligations from listed multilateral treaties are imported, they have been applied only *inter partes*, avoiding the creation of third-party rights. If non-ASEAN Member States were enabled to participate in an

ASEAN treaty, this might be facilitated through a non-Member's protocol that limited participation to priority matters for extra-regional cooperation, such as capacity building and intelligence cooperation.

In conclusion, the regional treaties to combat terrorism provide poor models. They are typically couched in vague language and contain many uncertain obligations. Their various definitions of terrorist acts and approaches to the criminalisation of those acts are conceptually flawed or inadequate. Most of their measures for prevention and intelligence cooperation are insubstantial. Their main strengths are in providing procedures for mutual assistance in investigations and in extradition arrangements. Nevertheless, the regional treaties do provide a framework for developing deeper intra-regional cooperation. Their aspirational goals address not only cooperation but also the building of national capacity to combat terrorism and those goals could be strengthened by measures to improve implementation. Therefore, despite the various inadequacies, they offer many lessons for the drafters of an ASEAN regional treaty.

How likely it is that ASEAN Members might actually adopt a regional treaty to combat terrorism? In recent years, ASEAN cooperation on counter-terrorism measures has moved rapidly forward at both the region-wide level and at sub-regional levels. The issue of terrorism was highlighted at the International Conference on Terrorism in Baguio City in the Philippines in 1996. Since then, an ASEAN-Japan Forum in Tokyo was held in May 1997 to establish a network for information exchange on combating terrorism. The *ASEAN Declaration on Transnational Crime* was promulgated at the first ASEAN Conference on Transnational Crime in Manila in 1997.

The Declaration aimed at examining the possibility of regional cooperation on criminal matters such as terrorism, drug trafficking and sea piracy, and included discussions on extradition. Following the Declaration, ASEAN countries established the ASEAN Ministerial Meeting on Transnational Crime (AMMTC), which gathers on a biennial basis and brings together ASEAN bodies such as the ASEAN Senior Officials on Drug Matters and the ASEAN Chiefs of National Police. The possibility of establishing an ASEAN Centre on Transnational Crime was entertained at the first AMMTC.

The second AMMTC in 1999 produced a *Plan of Action to Combat Transnational Crime*. The third AMMTC in 2001 endorsed the convening of an "Ad hoc Experts Group on the Work Programme to implement the ASEAN Plan of Action to Combat Transnational Crime." These developments set the basis for regional cooperation in combating transnational crime. Following the terrorist attacks on the USA on September 11, 2001, ASEAN cooperative efforts to combat transnational crime began to focus strongly on terrorism.

The seventh ASEAN Summit in Brunei in 2001 produced the *ASEAN Declaration on Joint Action to Counter Terrorism*. In May 2002 ASEAN organised the Special ASEAN Ministerial Meeting on Terrorism in Kuala Lumpur, which launched the *ASEAN Work Programme to Implement the ASEAN Plan to Combat Transnational Crime*. Concerning terrorism, the Work Programme includes information exchange, harmonisation of laws, intelligence sharing, coordinating law enforcement, training programs and the development of multilateral or bilateral legal agreements to facilitate arrest, prosecution, extradition and the like. It also provides for ASEAN members to work towards a regional convention to combat terrorism (item 6.2(e)).

A few months later in July 2002, the ASEAN Regional Forum issued a Statement on Measures Against Terrorist Financing, addressing such issues as freezing assets, implementing international standards, exchange of information, outreach, technical assistance and compliance and reporting. In June 2003 the Senior Official Meetings on Transnational Crime (SOMTC) endorsed a proposal to form Joint Terrorism Task Forces when an affected Member country seeks assistance in investigating terror incidents. In August 2003, the ASEAN Government Legal Officers' Programme conducted a meeting on counter-terrorism that discussed regional harmonisation of national laws and the prospects for a regional treaty.

On October 7, 2003, ASEAN Members declared in their Bali Concord II, the intention to form an ASEAN Security Community that, amongst other things, seeks to promote regional solidarity and cooperation in matters of security. However, there have been difficulties in implementing many of the intentions and commitments under the above declarations and plans at the region-wide level. The meetings did not address institutional capacity building or produce mechanisms to coordinate ASEAN bodies, such as the AMMTC and the SOMTC. For some ASEAN countries, the lack of intra-state coordination and issues of state sovereignty also inhibit implementation.

As noted by ASEAN at the UN Security Council Counter-Terrorism Committee Special Meeting in March 2003, there is still much work to do at the regional level. Due to ASEAN regional obstacles in counter-terrorism cooperation, the current pattern seems to be to cooperate at the sub-regional level within the ASEAN framework. For example, the *Agreement on Information Exchange and Establishment of Communication Procedures* was signed in 2002 in Kuala Lumpur between Indonesia, Malaysia and the Philippines, and later by Thailand in 2002 in a Declaration on Terrorism. The suggestion that a regional counter-terrorism treaty be developed was put forward by Indonesia at the ASEAN Government Legal Officers' Programme meeting in August 2003.

Despite the suggestion having already been adopted in 2002 by ASEAN ministers in their *Work Programme to*

Implement the ASEAN Plan to Combat Transnational Crime, the suggestion received a mixed reception. It seems likely to proceed at the sub-regional level. As the counter-terrorism capacities of ASEAN member States improve, a regional framework for related legal cooperation becomes necessary. The first ASEAN legal cooperation agreement was signed in late 2004 and will operate between those Member countries that ratify it.

However, it does not articulate a common definition of terrorist activities and its obligations are subject to national laws. ASEAN members are, in fact, currently strengthening their capacities to combat terrorism through partnerships with other countries. For example, the *ASEAN-United States of America Joint Declaration for Cooperation to Combat International Terrorism* was procured in 2002. The sixth ASEAN-China Summit in 2002 produced the *Joint Declaration of ASEAN and China on Cooperation in the Field of Non-Traditional Security Issues,* which has recently been developed into a Memorandum of Understanding between ASEAN and China. The fourteenth ASEAN-EU Ministerial Meeting produced a *Joint Declaration to Combat Terrorism* on January 23, 2003.

In January 2004 the AMMTC met with Japan, the Republic of Korea and China to discuss cooperation between them against transnational crime. The Regional Ministerial Meeting on Counter-Terrorism held in Bali in February 2004 was attended by ASEAN Foreign Ministers as well as those from Australia, Canada, China, Fiji, France, Germany, India, Japan, New Zealand, Papua New Guinea, South Korea, Russia, Timor-Leste, the UK, USA and EU. It agreed to establish an ad hoc working group of senior legal officials to report on the adequacy of regional legal frameworks for counter-terrorism cooperation and to identify areas for improvement of cooperation and assistance.

At national level, a Memorandum of Understanding on counter-terrorism cooperation between Australia and the Philippines was signed in March 2003 and another between Australia and Cambodia on June 18, 2003. The intense current activity in counter-terrorism partnerships is likely to produce some results. The institutional weakness of ASEAN and the particular political sensitivities posed by Islamic terrorism in the region suggest that a legal formula for regional counter-terrorism cooperation will not mature in the short term, however.

Yet we anticipate that within the medium term (5 years) a regional or sub-regional treaty on terrorism is likely to be adopted. An ASEAN regional treaty could promote counter-terrorism measures by putting in place institutional structures and decision making processes to promote cooperation, coordination, shared expertise and common legal approaches. An objective common legal definition of terrorism is readily available.

At its barest it is simply "serious violence committed by non- State actors, directed at civilians, and intended to coerce a society". It can be criminalised in national legislation without reference to pre-existing national crimes or a transnational component in the act. These are the pared down essentials of the 1999 treaties on terrorist financing and of the CIS and OAU. Elaborations might address preparations for and threats of violence, violence directed at social infrastructure, international institutions and at non-combatants. Due to the transnational nature of much terrorist activity in the ASEAN region, prevention cooperation and mutual assistance in enforcement measures at the international level are essential.

They include prevention, cooperation by establishing controls and information exchange at customs and immigration barriers, consultations on security measures for protected persons, strengthened regulation of stockpiling of weapons and dangerous goods, and intelligence coordination and contact points. Mutual assistance can be enhanced by extending measures for the collection of evidence, extradition of suspects, transfer of witnesses and the like. The earlier regional treaties on counter-terrorism are of limited use as models in these matters.

They tend to have a tightly constrained application and to focus on extradition, while also being framed in soft, vague language that is subjective in application. However, the latest regional treaty examined, the *1999 Convention on the Prevention and Combating of Terrorism of the Organization of African Unity*, does provide many useful articulations of counter-terrorism cooperative measures that are useful as models. ASEAN Members could take the opportunity to build a better framework for regional cooperation and coordination to combat terrorism than exists in other regional treaties by learning from the weaknesses in those treaties. For example, obligations to enact offences should not be circular but should harmonise legislative reforms. Procedures to extradite alleged perpetrators should not allow for exception based on asylum.

An ASEAN Conferences of Parties could be mandated to review prescribed reports on treaty implementation and to oversee the implementation of detailed action plans for technical capacity development, so as to give political impetus to a coordinated regional programme of capacity building. Other issues that could be addressed in the framework include the provision of safeguards for the exercise of civil and political rights and relationships with other regimes and non-Parties. There is within ASEAN a great deal of opportunity to build capacity for national action and for international cooperation. There are strengths, resources and precedents within its Member States and many resources and willing partners beyond. As these coalesce in the medium term, the region will produce the clear and just legal framework for counterterrorism cooperation that it needs.

9

Terrorism–News Media Relationship

Coverage of Terrorism by Media

Zaoui is an Algerian national who turned up at Auckland International Airport in early December 2002, and asked for political asylum. He was immediately imprisoned. Relying on classified NZSIS information, the Minister of Immigration has declared Zaoui a threat to national security and issued a Security Risk Certificate against him, allowing his continued detention. The Herald pointed to the just-released decision of the Refugee Status Appeals Authority, which had found no credible evidence that Zaoui was a terrorist.

The Herald said:

The authority says that much of the unclassified material the SIS supplied to it was not only unsourced, but apparently drawn from news reports and internet material of highly dubious content and purpose. No attempt, says the authority, was made 'to excise opinion from fact'. Effectively, the intelligence agency was happy to play fast and loose in its desire to connect Mr Zaoui to the Armed Islamic Group (GIA) and suggest a link to al-Qaeda. What the editorial didn't mention, though, was that the Appeals

Authority had also slammed the Herald for being just as sloppy and unfair.

First, the authority pointed out that whoever leaked Zaoui's name to the Herald was breaking the law, and by publishing it, the Herald may be endangering Zaoui's life. But that was just for starters. The authority also said the Herald was 'speculating wildly' on Zaoui's background and making the same sorts of simplistic linkages based on the same sorts of dodgy sources as the SIS was. This was how the authority described one Herald story, entitled, ironically. The report is an extraordinary list of sensationalist and largely inaccurate claims about the appellant culled from internet sources of the type we have already adversely commented on. It is hard to disagree.

The article said Zaoui 'is believed to be a terrorist on the run with links to sinister organizations'. It said his name ...is linked to terrorist cells that have carried out bombings, beheadings and throat slitting from Algeria to France. The name crops up in connection with Osama Bin Laden's suspected Southeast Asian army... It talked about Zaoui's 'terror activities' in the Algerian civil war, which it noted claimed 100,000 lives. It said 'various internet articles' described Zaoui as a leader of the brutal GIA, responsible for numerous gruesome assassinations. It referred to his death sentence in Algeria for supplying weapons to guerrillas from Europe. It linked him, in various shadowy ways, to bombings at Metro stations in France, a political assassination in Afghanistan, and even the September 11 attacks in the United States. It is hard to imagine anything more horrible you could say about a person.

Nowhere did the article mention that Zaoui was a religious leader, an elected member of Parliament ousted by a coup, and an outspoken opponent of violence. Nor that his convictions in Algeria were almost certainly political shams, and were handed down in his absence. The Appeal Authority exhaustively checked the various claims against

Zaoui and found them all to be bogus or unsubstantiated. The authority said poor Zaoui was the victim not only of a series of terribly unjust trials, but also of credulous news organizations, who uncritically recycled claims from other news organizations, including the state-captive Algerian media, who in turn were parroting the same few sources, most of whom were biased or crooked.

The emergence of such untruths in New Zealand lends weight to the appellant's contention that many of his problems over recent years stem from Western journalists' and officials' ignorance of Algeria and their preconceptions and fears about Islamic 'terrorists'. The authority pointed out that Zaoui has been an unwilling passenger on a merry-go-round of misinformation: Highly prejudicial misinformation concerning [Zaoui] quickly acquires the status of received 'facts' – a process reinforced by the diffusion and recycling of these 'facts' by the media/internet as well as between intelligence services and immigration and other officials in a range of countries.

For example, said the authority, Algeria's Le Matin immediately picked up on the Herald story, writing: This former head of the dissolved FIS is suspected of having links with Al Qaida. The list of Algerians arrested abroad during the inquiry on Al Qaida continues to increase. This time, the problem has struck deeply into the heart of FIS, of which a representative figure has been the subject of questioning in New Zealand where the department concerned is trying to verify the existence of links with Oussama Ben Laden's [sic] terrorist group.

The authority noted that the article incorporated information from an Agence France-Presse story, which in turn based its information on the Herald story. Le Matin also included material from a 2001 story published in Vancouver's Asian Post, which said: US intelligence sources told the Asian Post that Malaysia has come sharply into focus in America's new war against terrorism. A previously unidentified group called FIDA or Sacrifice is currently being investigated in Malaysia.

Police sources told the Post that the coordinator for FIDA in Malaysia is a man called Ahmed Zaoui and that he works closely with FIDA affiliates in Sweden and the United States. The authority discredited the Asian Post report, finding that authorities in Malaysia, the United States and Australia were well aware of Zaoui's presence in Malaysia at the time, but none of them had made any move against him; and that Interpol Washington confirmed in December 2002 that he was not wanted by the United States.

Yet these allegations were still being given currency by the Herald – apparently without any attempt to check them out – in July 2003: Media reports from Vancouver in 2001 linked Zaoui to Osama bin Laden's secret army in Southeast Asia. Two weeks later, the Herald found itself able to report, with stunning disingenuity, that 'since he arrived in New Zealand last December, Ahmed Zaoui has been portrayed as one of the most dangerous terrorists on earth.' The article drew on the authority's findings to debunk that portrayal. But nowhere did it mention the paper's own role in perpetuating it.

The Herald seldom gave the reader the benefit of its sources, even in a general way, simply calling Zaoui 'a member of a militant Islamic group responsible for a range of atrocities and terrorist activities in Algeria and in Europe'; or saying he was 'linked to terrorist acts' or was 'alleged to be a member of the Armed Islamic Group, accused of atrocities against Algerian citizens and acts of terrorism in Europe'.

In particular, the Herald gave no hint that any of its sources may have had their own interests to peddle in portraying Zaoui as a dangerous criminal. The authority had found: that the Algerian media was tightly controlled by its government, which had long been hostile to Zaoui, and had to be treated with caution; that many Algerian news releases were recycled uncritically in the French press; that the information in circulation came from a 'very limited number of original sources'; and that many of those sources were 'suspect or dubious'.

It criticised the research of the SIS, saying it 'largely consisted of searching the internet for "hits" on [Zaoui's] name'. It seems that the Herald's approach was little different. To be fair, as early as 16 December 2002, the Herald was raising doubts about Zaoui's terrorist connections, quoting an Islamic leader calling Zaoui 'a peaceful person in no way linked to terrorist activities' and a terrorism expert saying it was possible Zaoui had been wrongly accused of terrorism.

And in the middle of its lengthy summary of the Appeal Authority's decision, the Herald did note that 'the report criticises New Zealand media – including the Herald – for relying uncritically on foreign newspaper and internet reports about Zaoui'. But the thrust of its coverage, and particularly in the first few weeks after Zaoui's arrival, was the constant repetition of serious (but largely unsubstantiated) allegations of his deep involvement in terrorist activities. The Herald was not the only culprit.

The rest of the media leapt on the bandwagon too. For instance, the Waikato Times and the Christchurch Press both published a NZ Press Association story that said Zaoui's name was 'linked with Osama bin Laden's terror network in Southeast Asia'. Welling- ton's Dominion Post reported that Zaoui's name 'is the name of an internationally wanted Algerian terrorist who was a leader of the Algerian Islamic Salvation Front, blamed for massacres in the 1990s' and 'was linked by Canadian media reports to Osama bin Laden's secret army in Southeast Asia.' But when the Appeal Authority debunked these claims, none of the other papers – with the honourable exception of the Press – made anything of the authority's rebuke of the media in their reports of its decision.

How could the media be so sloppy? I suspect they got caught up in the sort of terrorism-inspired alarmist fervour that saw the Dominion Post lead last November with news that an Air New Zealand flight from Brisbane was hit by a security scare when ...well, a woman overheard something threatening- sounding, that may or may not have been a joke, from someone who may or may not have been aboard the flight, that in any event arrived safely. A week earlier, the Truth had warned that Muslims may mount a large-scale terrorist attack on Sydney – and other major cities – maybe during the World Cup final, based on ... well, a couple of websites warning Muslim brothers to leave major cities because large-scale terrorist attacks were planned.

In October, the Press went big with a dispute over the attempted purchase of a Christchurch mosque by a Muslim charity that... well, may have two branches out of fifty that gave money to organizations connected with al Qaeda, though as a whole was considered legitimate by the US and New Zealand governments. So the Zaoui coverage should perhaps be seen in the light of the culture of terrorist hysteria pervading the news media these days. Moreover, it has been heartening to see the more recent coverage of the Zaoui case (he has won a succession of court cases against the Government as he challenges his continued detention) become more balanced as the media zeroes in on the quality of the evidence – or lack of it – against Zaoui. This is where the focus should have been from the start.

The media, in their own way, have the power to damage Zaoui as much as the SIS when they play fast and loose with facts. It was a story that seemed to conclude that European intelligence sources agreed there was no evidence that Zaoui had ever engaged in terrorist activities, but still believed he may have supported terrorists. No comment from Zaoui was included.

The news media are an extremely powerful actor in the dynamics of oppositional political terrorism. Just how important newspapers, radio, and television are during ongoing campaigns and in the context of terrorist incidents is a subject of constant debate. Any understanding of the connections between this type of violence and the media must be embedded in broader discussions of: the power of the media, especially in conflict situations; the relationship among journalists, editors, authorities, and terrorists; empirical analyses of the media; and the connection between terrorism and public opinion.

Nevertheless, since the early 1970s, researchers have examined the role of the news media in connection with terrorism. In many respects, this body of work is a sub-specialty in the field of terrorism studies. It is typically anchored in a limited number of academic disciplines including communications studies, law, political science, and sociology. In short, this Research Note looks at the venues in which research on this connection is typically found, the topics that academics generally research and the methods they use, their findings, and where these scholars might devote their future energy.

The majority of scholarly research on the connection between terrorism and the news media has appeared in the form of articles in peer-reviewed journals or chapters in scholarly books. Less frequent are stand-alone monographs. A considerable amount of the books on the terrorism–news media relationship are edited collections, typically consisting of papers originally presented at a conference sponsored by an academic organization or think-tank. Other pieces are part of edited books that examine the news media's role regarding violent political conflict. The language, with which the media reports and discusses insurgent terrorist organisations and their actions, is extremely important, as the language which it adopts often will set the parameters for public discourse.

The phraseology and terminology of the insurgent terrorists and government officials are generally at odds, thus the media is forced to adopt one or the other's word or phase which, in turn, will generally become the accepted way to express that idea in the public forum. Therefore, if the terrorist organisation or the counter-terrorist group can induce the media to accept their nomenclature, it has already won an important psychological victory.

Most studies into the relationship between terrorism and the media have focused on the response of the media to terrorist incidents. They have generally agreed that the relationship between terrorism and the mass media is 'symbiotic', in that insurgent terrorist organisations use the media as a conduit for their political message to be heard by the target audience, whilst supplying 'exciting news' for the media. Furthermore, most previous studies have focused on either the political, legal or psychological aspects of the relationship between the media and terrorism, while largely ignoring other methods of analysis.

This work aims at taking the first tentative steps of expanding the study of the relationship between terrorism and the media by taking a linguistical and semantical approach. To achieve this aim, this text will be broken into two distinct sections. The first will be a brief discussion of the ways in which the terrorist's and counter-terrorist's language can make its way into common usage in the media. The second section will introduce some of the semantic issues that arise from value and moralist language becoming the acceptable language in public discourse. In Janny de Graaf's esteemed text, Violence as Communication, de Graaf argues that when journalists interview sources there is a 'good chance' that they will also inadvertently adopt some of the source's language; that means in practice, that when a journalist uses an insurgent terrorist as a source, the terrorist's romantic language often seduces the journalist into unconsciously adopting it.

An example of this phenomenon occurred during the kidnapping and subsequent murder of former-Italian Premier Aldo Moro, when the editor of La Repubblica ran the headline 'They Have Struck The Heart of The State', which seems to be a direct paraphrase of the Red Brigade's statement '...we have carried the attack into the very heart of the state.' The terrorist organisation had clearly excited the newspaper with their engaging language.

The media, however, does not only adopt the language of the terrorist. De Graaf also pointed out that 'in many cases' the news media automatically adopts the nomenclature of the government. However, most commentators do not allege that the media is seduced by the language of the government, rather is intimidated by the government's perceived information superiority. Moreover, due to terrorism's enormous emotional impact, there is often a lack of neutral words with which to describe the incident. There are, for example, few neutral nouns for journalists to describe an insurgent terrorist, as, 'terrorist', 'soldier', 'freedom fighter', 'criminal', or 'guerrilla' all require the journalist to make a moral judgement.

Therefore, often journalists are forced to employ words which seem to indicate a bias out of lack of a more neutral substitute. Thus, whichever terminology the media adopts quickly becomes, in Schlesinger's apt words, the 'primary definitions of social reality'. Take for example, John Mold's comments in the 'Letters' section of The Australian, ...George Bush... (and)...neo-conservative thinkers like Paul Wolfowitz, whose aim appears to be the protection of Israel under the guise of a war against terrorism.

The Bush/Wolfowitz policy of pre-emptive strikes against alleged rogue states starting with Iraq, is a sure path to continuous war which should not involve us. Our only hope for world peace is the UN and continued and increased support for the body. (Emphasis added) It is clear that Mold disagrees with the current Australian policy, however in order to join the debate Mold had no choice but to employ the United States Government's anti-terrorist rhetoric, such as 'pre-emptive strike' and 'rogue state'.

The media, by adopting the United States' anti-terrorist vocabulary, has limited the way in which Mold can express his views, thus has successfully set the parameters of the public debate. George Orwell dedicated much of his academic career to highlighting the threat to an individual's cognitive processes from what he disparaging termed 'journalese' or 'officialese'. Orwell argued in Politics and the English Language, that when an individual becomes a slave to official jargon they are, in a sense, gagged. Individuals are prone to use 'officialese' and follow the 'mindless thought grooves' which, in Orwell's opinion, could easily be replaced with more precious and thoughtful terms.

However, 'officialese' is continually regurgitated by citizens in the public discourse without any knowledge of the semantic meaning of the language that they employ. Orwell was a student of semantics, and thus fully knew of the close relationship between language and thought. Although the validity of the concept of 'linguistic determinism' is an ongoing debate within the social sciences, it has been well established that the 'pre-packaging' of language has a direct impact on thought. The 'pre-packaging', or the prior digestion, of concepts and ideas generally results in an oversimplification and stereotyping of language.

From this perspective, one can see that one of the functions of the media is to 'pre-pack' terrorist incidents and make them and increase the easy in which an audience can assimilate them. For example, terms such as 'terrorist', 'act of terror', 'fundamentalism' or 'threat' act as familiar signposts for the audience, allowing them to give structure to their thoughts, which would otherwise be incomprehensible due their complexity. In short, the concepts of black (terrorist?) and white (counter-terrorist?) are relative easy to describe, as they are bi-polar, however shades of grey (where truth and reality normally reside) is notoriously difficult to describe in words.

Therefore, in conclusion, words are more than mere symbols that convey meaning. Words influence thought and limit the ideas and concepts that can be transferred from one individual to the next. As we have seen in this paper, the media plays a central role in telling the public what words will be judged by society to be appropriate in any new discourse. Thus, it is of crucial importance to both insurgent terrorists and state agencies that the media uses their language to describe acts of politically motivated violence.

Terrorism and the Mass Media after Al Qaeda

Marshall McLuham, one of the most celebrated researchers on the social impact of the mass media, came to the relatively precocious conclusion that "without communication terrorism would not exist." This short sentence holds a truth that has remained unchanged during the last decades. It is not uncommon to cite the relationship between the diffusion of terrorist messages and the existence of modern mass media. Terrorism, however, did widely exist before the mass media did. An example of this is the type of terrorism that promoted anarchy (an authentic plague for nineteenth-century societies) utilizing assassinations and other types of attacks as a way to reach entire societies.

The killing of important figures or other actions that took place in front of hundreds or thousands of witnesses, were effective means of ensuring that these events were made known during a time in which there was strong governmental control over information and the mass media acted within a limited scope. In the future, technological developments would allow terrorist attacks to be published in ways that were previously unimaginable. Terrorists found a powerful ally in the mass media that would help them obtain public attention for the group and its demands.

According to Brigitte Nacos, one of the scholars that has most studied this issue, terrorists commit violent acts looking for three universal objectives: to get attention, to gain recognition, and even in order to obtain a certain degree of respect and legitimacy. These objectives are attainable for those individuals that are capable of receiving the most media coverage. And those that obtain it have more opportunities to influence others. Terrorists always calculate the effect that their actions will have in the media and the overall probability that this will provide them with the opportunity to be a member of the "triangle of political communication".

This image exposes one of the principal characteristics of contemporary society where personal and direct contact between the government and the citizens is no longer possible; a situation where the mass media is in charge of providing the channels of communication between those that govern and their constituents. Access to the media brings terrorists closer to a democratic society's decision-making process, which greatly increases that chances that this complex network of interactions will result in a political decision that favors the interests of their group. The objective of this article is to more closely examine the type of relationships that can be established between a terrorist group and the news media.

As part of our case study, we will utilize the terrorist organization Al Qaeda, given that with this group it is possible to encounter diverse models exemplifying its relationship with mass media. Our methodology will involve analyzing the content of its public statements and examining the developments that have taken place during its history as an organization. Both perspectives suggest that terrorism's view of the media, far from being composed of rigorous ideological or political principles, is shaped by their calculations of estimated opportunities.

Modern Terrorism and TV Logic

Any study of the relationship between the media and terrorism would be incomplete if it did not mention the peculiarities of the principal and most influential medium of mass communication. In fact, television's consolidation as the principal source of information and knowledge for millions of people meant a new step in the evolution of the terrorist phenomenon. TV has a series of characteristics that make it easily adaptable to terrorist logic creating a situation of almost perfect "symbiosis": The TV news bulletins are focused mainly on what is known as "visual culture."

The attention that a given news event receives is directly proportional to the amount of audiovisual material available for it. Many news bulletins are extremely adverse to covering and presenting a story using only narrative o audio components. The availability or lack of audiovisual material becomes a determining factor when choosing which stories will be included in the broadcast and which ones will be excluded. This principal trait of mass media has repercussions on the planning of any possible attack. As a result, terrorists look to attack those places that given their location or their significance will attract the immediate attention of the media. This, in turn, provides a sufficient quantity of images that will guarantee one's presence on the news.

Following this logic, the example of the September 11, 2001, attacks were sufficiently visual to meet the demands of the TV culture and to satisfy the public fascination for live coverage of events. By attacking cities like New York, among others, where the largest concentration of television stations and film studios and equipment exist, terrorists not only guaranteed for themselves an exhaustive coverage and a global projection of their actions, but the existence of multiple tourists and citizens who had their own film equipment. This allowed news broadcasters to utilize domestically filmed new materials in which the news did not have as much to do with the terrorist attack as with the existence of new images that allowed the viewing public to contemplate the horror and destruction from a different point of view.

Time on TV, by definition, is short. Any topic covered on a televised news report is subject to strict boundaries of time limitation. The search for a compact format, where one can summarize and offer the viewer the principal news events in a brief time period make deep analysis, knowledge of historical context, antecedents and any other element which escapes the realm of the immediate highly difficult. This constitutes an enormous limitation when informing citizens about a series of issues that deeply affect their interests. The fact that that the majority's perception of terrorism is of audiovisual origin, where clichés predominate alongside of simplifications and shallowness, has an enormous influence in the way in which public opinion presents its demands to the public officials.

This influence is also seen in the latitude with which these political decision makers can operate. Time also constitutes a problem, not only because of its shortage, but also because of its poor use. Mass media is strongly dependent on novel aspects that create a new "story." They look for attention-drawing elements and let them to stretch out a story as much as possible, making less significant events, big news flashes with the purpose of slowing down events while waiting for new headlines. The TV prioritizes violence indirectly. The media selects events that are news worthy (based on its own set of values that give priority to violence and conflict in any form).

The broadcasting of a news program involves uncountable preparatory actions that involve the selection and discrimination of content that daily is placed on the desk of any entity that works and write stories that will come out in the mass media. There is competition among the different stories that will finally be emitted; those that are victorious are more dramatic, are more spectacular in a visual sense, are more emotional, and contain other elements that are able to be assimilated by an image-oriented culture.

The result is not only the shadowing of those events that, despite their interest, lack a conflicting nature, but the establishing of a dangerous pattern for those that want to be "made public" at any price. On the other hand, different studies demonstrate that the visual presentation of violence and brutality by the news media creates feelings of fear among individuals that are not directly exposed or who have not suffered said actions. In fact, televised coverage of a terrorist attack's effects (especially if it is live) creates a paradoxical situation in which the spectators imagine more horrendous scenes that the very witnesses situated in the area . Terrorism's permanent and unconditional presence in current mass media "overdimensionalizes" its capacity for reaching the population, strengthening the effects of its threats and coercion.

The media's capacity for concentrating and maintaining its attention over a determined issue is weak. Events normally stop being discussed in current news stories when they can no longer repeat those elements that attract the mass media's attention. A never-ending search for novelty causes the media to leave out stories that, despite having initially received widespread attention and regardless of their continued importance, are suppressed or put aside. The reasons for this are the supposed saturation that could be produced in the viewing public after watching the "same images" day after day. The fact that the mass media search constantly for novelty and new points of view makes it, from the terrorists' perspective, a "fickle friend."

As a consequence, they attempt to perform violent actions in rhythmic succession, or innovate in a way that allows them to regain the media's "favor." Television attempts to promote a self-image of truthfulness. Stations do not hesitate to present the contents of the stories that they broadcast as authentic "pieces of reality," exploiting the credibility that one usually gives to anything that "can be seen". The relationship between the credibility that one gives to what he or she perceives visually and this use of the trust factor by the media has been an object of reflection for Italian political scientist Giovanni Sartori.

According to him, the media distorts the knowledge gaining exercise because it does not remind its viewers that the information that an image can "transmit" is a question of framing, the use of a particular camera, the choice of one focus over another, etc. The inherent truthfulness that one assumes exists in everything that "can be seen" gives television a persuasiveness that other media lack. This signifies that the product that TV stations offer is "reality in and of itself," without taking into account the context and a group of other circumstances that "surround" these images. Given this visual advantage, television does not hesitate to claim its capacity to reach the news story wherever it may take place. This proclamation, however, holds a dangerous misconception.

Financial limitations cause the media to concentrate its resources where "it is believed" that there is higher news interest. This leads to a situation in which their interests in every corner of the planet have more to do with determining beforehand where a certain news event "can" take place than with where news is really happening. Thus the mass media itself generates newsworthy events, devoting a disproportional amount of attention to frivolous and irrelevant occurrences, excluding other situations that can affect tragically thousands of people causing them to be simply be forgotten or of anonymous character. The result of the aforementioned situation is the existence of several corners of the planet that are never discussed in the news, despite the emergence of problematic situations that could affect the world as a whole.

This asymmetric quality of the news causes terrorists to concentrate their efforts in those places where they can receive the media's attention, setting aside other places where their actions receive no interest aside from the violence itself. Searching for a more human, intimate, and accessible side of current events, TV shows a strong tendency to "personify" the story. The media's attempts to catch the viewer's attention by giving a face and human qualities to problems whose genesis and complexity completely exceed the limitations of any one individual. This way of focusing on reality has repercussions on the way in which citizens face problem formulation, setting aside the collective, cultural, transnational, and ideological aspects that underline the vast majority of terrorist movements.

Even when it is tremendously difficult to personify a problem in one or several concrete subjects, the news tends to perform this very operation utilizing a series of stereotypes of clichés that are widely diffused among the population. Following this pattern, one frequently finds information that rests on popular clichés and cinematographic imagery, when offering information on terrorists, mafias, secret agents, etc. These types of operations help to favor the appearance of "charismatic celebrities".

Mass media's weaknesses and limitations when focusing on the terrorist phenomenon are not only found in "structural" elements, as one might call them; they also feed on a series of ideological conditioners knows as "focuses." We can understand a focus as the active construction, selection and structuring of information that the news media carries out in order to fill a particular reality with meaning for the viewer. This type of focusing takes place when the mass media highlights determined aspects of a particular issue making them more preeminent, promoting one definition of a particular problem or a moral judgment, or giving a recommendation on how to resolve a troublesome issue. The media are not a simple entities that transmit "raw" information.

It has an active role in "constructing" the news, making terrorists actions understood in a context that simplifies, prioritizes, and structures the narrative flow of these stories. The fact that popular perception on terrorist violence is shaped, to a large extent, by mass media's ideology and the underlying moral point of view is not necessarily negative. Presenting terrorism in a manner that clearly rejects violence, demystifies assassins and emphasizes victims' personal tragedies is an essential resource in any type of multi-faceted strategy that combats terror.

The problem arises when media controlled by certain ideological, cultural or religious conniving begins to look for a fictitious balance between murderers and victims; when the "causes" of certain terrorist groups are justified against others; or when a story is treated in such a way that it gives the viewer the impression that terrorist groups are political participants that deserve certain legitimacy in the competition for power.

Relationship between Terrorists and the Media

The majority of a considerable amount of academic literature where this topic has been discussed has pointed out the symbiotic relationship between terrorism and the mass media. In other words, it is the type of relationship between two groups involving mutual dependence where one party complements the other. There is, however, more than one type of relationship that can exist between terrorists and the mass media. The diversity of causes, ideologies, and social and cultural conditioning factors that inspire different terrorist groups are equally applicable to the type of relationships that these groups establish with the news media.

Michael Wievioska, a French sociologist, was one of the first specialists to dissect this plural nature. He would establish four different models of behavior: Complete indifference. The terrorists' goal is to terrorize their victims, without seeking to attain media attention for their acts. There is no expectation that the press will become involved. This French author does not hesitate to mention that this situation is highly unusual.

Relative Indifference

The terrorists are not concerned with being on the news, even though they are conscious of the power that explaining their cause in currently existing media can provide them.

Media-oriented Strategy

The terrorists are not only aware that the press can expand the scope of their words and actions, they also perform a series of operations based on the knowledge that they possess on the dynamics and functioning and the of news producers. After well thought manipulations, the news media becomes integrated in the terrorist group's actions. Complete breakaway. This is case of terrorists that see journalists and reporters as enemies that must be destroyed, putting them on the same level as other direct adversaries.

The press ceases to be an entity that should be cynically manipulated (as democracy's Achilles' heel). It is instead viewed as the appendix of a system that must be destroyed. These four options do not only give us an idea of the variety of possible relationships that can be established between these two entities, it also allows us to see how terrorists can alternate between different models, or even adopt more than one at the same time. In fact, modern terrorist groups frequently employ strategies that work to satisfy the news media's demands, while at the same time considering them an integral part of an enemy that they seek to overcome and annihilate.

Jihadist Perception of the Media

A large part of the structural characteristics and the modus operandi of Al Qaeda are novel within the general history of terrorism; however, its relationship with mass media has long followed the common patterns known up until now. In fact, this attitude falls into the last two categories that Michael Wievioska formulated. On one hand, the group is aware that its final goals are unreachable without the wide and constant presence of the organization in the press; however, on the other hand, its messages brim with hostility toward those that they consider to be guilty of the failure of the Jihadist movement.

From Al Qaeda's point of view, the news media are principally responsible for the liberating message of the Salafist Islam being ignored or distorted. This makes it impossible for the Jihad to penetrate in large sectors of the Muslim community, which finds itself immersed in the most pure ignorance and error: "If we add to the foregoing the media siege imposed on the message of the jihad movement as well as the campaign of deception mounted by the government media we should realize the extent of the gap in understanding between the jihad movement and the common people."

The media are the principal authors of the stereotyped and clearly negative vision that society has of the organization's participants and activities. The press is considered to be mere extensions of the enemy. Along these lines, Muadh bin Abdullah Al-Madani, Al Qaeda ideologist and the author of one of the most extended jihadist articles following the September 11 attacks, expressed it in this way: "This past year has seen the media, which pleads impartiality, churn out their most awesome propaganda machine bombarding the watching public with the White House spin on events, completely ignoring any other perspectives.

In all the reams of newspaper articles, stacks of video

footage and radio interviews, has there been a single dedicated documentary into the 'hijackers' and their motives for the attack? No. Why? Because it is more convenient to label them as madmen, crazed, psychotic evil doers who either wallowed in the squalor of abject poverty and therefore had nothing to live for, or were psychotic fanatical killers easily brain-washed by power hungry religious nuts."

Ayman Al Zawahiri, number two in Al Qaeda, enumerates the "tools" that the "Western powers" use to fight against Islam in his book "Knights under the Prophet's Banner". In this list, "international news agencies and satellite television" has a very prominent role, together with other entities like the United Nations, western companies, non religious governments, and international aide organizations. Jihadist propaganda frequently denounces the pernicious role that the media play in the covering the "war" that the Muslim world is carrying out.

According to Al Qaeda, the press continuously lies, not only about the real motivations of the mujahidin, but above all about the real motivation of Western governments. These are presented on the news as entities that look to find peace and international security, hiding the economic exploitation of the Muslim world and a deep rooted hostility against the Islamic religion and its followers. The Jihadist vision of media's role and their influence in the development of this "new crusade" can be resumed in the following points: The media numbs the population in general and particularly Muslims, entertaining them with futile and sinful issues. Their goal is to keep Muslims unaware of the seriousness of their situation as a people and of their obligations as believers: "The media sector is in the same category as it strives to beatify the persons of the leaders, to drowse the community, and to fulfill the plans of the enemies through keeping the people occupied with the minor matters, and to stir their emotions and desires until corruption becomes widespread amongst the believers."

They contribute to creating a false image of Islam's enemies' strength. Jihadists are firmly convinced that American prowess and influence a have been constructed over a series of mistruths diffused by the media. They have taken charge of creating the image of an invincible superpower, magnifying its military capacity to the point of delirium. This, in turn, has made it possible that many decide not to rise against the injustice due to the fact that they are convinced that they will be defeated: "Oh people, do not fear America and its army. By Allah, we have struck them time and time again, and they have been defeated time after time. In combat they are the most cowardly of people. Our defense and our war against the American enemy have shown that [America's] warfare is mainly psychological in nature, because of the vast propaganda apparatus at its disposal."

"The American battle is a psychological battle that depends on the media and the magical effect of the microphone. This campaign was successful in some of respects in Afghanistan, due to the absence of the counter psychological warfare in all of its forms." The media distorts the Islamic combatants' true colors and motivations. Said actions prevent the population from adequately appreciating and understanding the mujahidins' actions.

The chances for creating widespread revolt in the Muslim world are weighted down by the mass media's pernicious filter: "The Arab and Western media are responsible for distorting the image of the Arab Afghans by portraying them as obsessed half-mad people who have rebelled against the United States that once trained and financed them the purpose of the distortion campaign against the Arab Afghans is clear and obvious, namely, the wish of the United States to deprive the Muslim nation of the honor of heroism and to pretend to be saying: Those whom you consider heroes are actually my creation and my mercenaries who rebelled against me when I stopped backing them."

The media contributes to the utilization of aggression against Islam because it ensures a sufficient amount of social support for the policies of Western governments. Using these methods, the press takes responsibility for extending hate toward Muslims and gaining the viewing public's acceptance for the use of violence against them and the occupation of their land: "There are many innocent and good-hearted people in the West. American media instigates them against Muslims. However, some good-hearted people are protesting against American attacks because human nature abhors injustice."

"As for the decisions made by Bush and the US administration to prevent satellite channels and world news agencies from making our voice heard in the world, then this is clear evidence that the US administration fears the revelation of the truth that led to the Tuesday events." Terrorists, however, are not immune to the enormous power of the mass media.

This leads to a contradictory dependence on them. Jihadists are aware that they are fighting a battle where main setting is the news media itself. Ayman Al Zawahiri recognizes this in a personal letter: "However, despite all of this, I say to you: that we are in a battle, and that more than half of this battle is taking place in the battlefield of the media. And that we are in a media battle in a race for the hearts and minds of our Umma. And that however far our capabilities reach, they will never be equal to one thousandth of the capabilities of the kingdom of Satan."

Al Qaeda places enormous value on the benefits generated by continuously forming part of the evening news. The news media help to multiply terrorist groups' power. Their mere appearance in news bulletins allow terrorists to strengthen the psychological effects of their actions, to call society's attention to their message, and to offer a favorable image of their power and capacity. In one way, they are cynical of the press considering it to be an enemy against whom they should fight, but in another way it is seen as a valuable resource which the terrorist organization can utilize in order to achieve its goals.

A good part of Al Qaeda's attitude toward the media can be explained by some of the organization's experiences during the second part of the 1990's. The "Declaration of War against the United States" in 1996, for example, caused little impact. With this communiqué, Bin Laden hoped to carry the banner of Islam's battle against "Crusaders and Jews;" however, his declaration only caused commotion in radical sectors, and went practically unnoticed in the country on which war was declared. Unwilling to commit the same error, Bin Laden spend two year undertaking on a publication campaign in the news media and gave various interviews with Arabic and Western journalists before reaffirming his threats against the West in 1998 this time under the umbrella of "World Islamic Front for Jihad Against Jews and Crusaders".

This openness to the media had a series of immediate effects for the group such as an increase in donations, new recruits, and the growing and increased psychological impact of their actions. Journalist, Jason Burke, in one of his books discusses that in early 1998, Osama Bin Laden sent a signed letter to a collaborator in Pakistan in which he asked that certain journalists be paid more; his goal was to increase coverage of his statements and activities. His interest in gaining notoriety and fascination for power in the media arrived at such heights that some of his collaborators consider that Osama "has caught the disease of screens, flashes, fans, and applause."

He contemplates how the international media greatly contributes to Al Qaeda reaching their ultimate goal of becoming an ideological reference. For this reason, the organization started to openly ignore the Taliban's requirements of a covert operating style. Following the Americans' invasion of Afghanistan, renown Jihadists, like Abu Walid al Misri, member Al Qaeda's Shura Council did not hesitate to blame Bin Laden's obsession for appearing on the news for the "disaster" that losing Afghanistan meant for Islamism: "At that time, Bin Laden was obsessed with the media, the international media in particular. Mullah Omer could not restrain Bin Laden's words. However, Bin Laden was prepared to sacrifice Afghanistan and Mullah Omer, in exchange for making his statements."

Turning Point

Al Qaeda's negative perception of the mass media has been largely conditioned by the Western media's leadership in the world and by the airtight political control of the press in the Arabic-Muslim world. In both cases, Al Qaeda detects a deliberate interest in silencing the mujahedeen or distorting their image. It does not aspire toward its message receiving any type of understanding, given that this would imply questioning the legitimacy of political regimes that are "protected" by the media's actions.

The terrorist group is reconsidering, however, their Manichaean view of the mass media with the emergence of an outlet that is radically transforming the audiovisual panorama in the Islamic world: the satellite television station Al Jazeera. Many individuals have written about the controversial relationship between the terrorist organization and this TV station. In fact, Bin Laden's organization has been particularly partial to this Arabic channel, using it to gain publicity for his most important post-9/11 manifestos. This station has also enjoyed some exclusive privileges as far as the sending of propagandistic material and conducting interviews.

Along these lines, for example, this station was the only one that was able to interview Bin Laden after the attacks in Washington D.C. and New York. The interviewer was journalist Tayseer Alouny who was arrested and prosecuted in Spain for his membership in Al Qaeda. Several data like the fact the Bin Laden himself confessed to being a "regular viewer" of the TV channel or that a dozen members of Al Jazeera have been arrested with charges of collaboration with terrorism combined with the directors' refusal to facilitate information on how the network is able to make contact with Al Qaeda and their ability to constantly obtain previously unpublished information leads detractors to denounce the existence of a true ideological meeting of the minds between the Bin Laden network and the Qatari station. In order to address these doubts, it is necessary to thoroughly examine the TV channel's genesis in a wider context.

The monarchy that governs Qatar created Al Jazeera with the goal of multiplying this diminutive emirate's influence and international presence. If we take in account the way that this small kingdom (with a population 863,051 people according to data collected in 2003) has made a place for itself channeling issues that affect hundreds of millions of Muslims, we can classify this initiative as being one of the best investments in history. In fact, Al Jazeera is considered a paradigm of "asymmetric interdependence," given the disproportionate influence and impact that a broadcasting network can have on international affairs and on public opinion with respect to the miniscule quantity of political power that Qatar as a country possesses.

Given that the rest of Arabic TV stations, strongly controlled by their governments of origin, are known for their lack of credibility, Al Jazeera has become the preferred network for the majority of Muslims, regardless of the country where they live, their social situation, their level of religiosity, and their political preferences. It has a varied staff coming from different countries helping to place the foundations for a Pan-Arabic identity; this is reflected in its wide editorial coverage. All of this is enhanced by attractive visual surroundings and narrative techniques "imported" from U.S. news bulletins. Al Jazeera began its consolidation covering the so-called "Second Palestinian Intifada" in 2000. Its ample and graphic manner of filming the confrontations, sparked a wide series of pro-Palestinian demonstrations throughout the Middle East. When the network broadcasted Arabic citizens' opinions calling for their leaders to do more for the Palestinians, governments of the region quickly reacted accusing the broadcaster of inciting violence. Several Arabic governments, including Egypt and Jordan, declared that Al Jazeera's coverage of the insurrection threatened their regimes' stability and exposed them to their own people's criticism. In fact, Egypt and Jordan have been more critical of Al Jazeera then even Israel.

The continual criticism that the network has received since its inception from political regimens in the Islamic world has been one the key reasons behind its popularity and acceptance. Tight government control of the domestic media has praised Al Jazeera's role as a pillar of free expression in the region. For the first time, Muslim audiences can feel identified with a channel that not only broadcasts in their religion's language but also does this without fearing those in power.

Al Jazeera, however, has not only been the first broadcaster that openly criticizes powerful individuals, it has also been completely revolutionary in the way it treats certain issues that are of interest to the Arabic world. Despite the fact that its editorial line is clearly pro-Palestinian, the broadcaster has no qualms about feeding the controversy by including in its stories "the other side's point of view." The Qatari station does not only interview Israeli officials, it also discusses certain unquestionable ideas in the Islamic world. Al Jazeera reached maturity starting with the coverage of the September 11 attacks. According to one member of the network, it is estimated that subscriptions via satellite increased 300% during the month that followed 9/11.

The human tragedy of these terrorist attacks unleashed a wave of solidarity toward the United States in the international media. This includes some traditionally anti-American Arabic channels that decided to take a break before recommencing an editorial line fomenting hate toward the United States and Israel. In these moments, it was possible to find disaccord in Al Jazeera, which allowed them to increase their audience's loyalty. The Qatari network's programs not only sheltered the most outlandish theories pointing toward an American and Jewish conspiracy whose objective was to blame the Arabs, they also had no qualms about being spokespeople for Bin Laden and his organization's messages.

This was not the first time that Al Qaeda used Al Jazeera's signals to transmit its messages. The network had previously interviewed Bin Laden who had been the object of wide news coverage for years making the Saudi Arabian relatively well known to Al Jazeera's habitual viewers; the opposite was true in the Western world, where Bin Laden was largely unknown. Broadcasting this organization's post-9/11 videos and manifestos was a risky gamble for the network, given that this treatment meant making the U.S.A. and Bin Laden moral equivalents. The channel did not only take charge of extending and amplifying these messages, it also left room in its broadcasts for individuals that did not hesitate to zealously defend Al Qaeda and the need for a jihad against the West.

This support was positively valued and met with gratitude by the terrorist organization that responded by rewarding the network with some of these years' most important exclusives. Al Qaeda found the network to be an important and efficient spokesperson, a broadcaster that continuously validated its message due to the way it covered given news stories.

The organization's discourse emphasized the opinion that the American invasions of Afghanistan and Iraq were genocidal and viscerally anti-Muslim in character this being the very type of news that a viewer could find on this TV station. Al Jazeera does not hesitate to highlight the most human and emotional side of the Muslim victims of these conflicts, presenting the insurgent and terrorist groups that attack Western troops as legitimate resistance to unjustified aggressions. The beneficial perspective for the jihadist discourse allowed Al Jazeera to be the only network present in the Taliban territory during the U.S. invasion. The network played an important role in delegitimizing the American response to the 9/11 attacks.

The impact that their chronicles on the civilian victims of the conflict had on the world's public opinion combined with the supposedly deliberate nature of these casualties increased the American administration's anger who did not hesitate to "mistakenly" bomb Al Jazeera headquarters in Kabul. This belligerent way handling of information has also taken place during the U.S. intervention and occupation of Iraq. Its qualifying of Iraqi terrorists and insurgents as "resistance" has led the new Iraqi government (strongly hit by these groups' actions) to temporarily suspend the network's activities in the country and not grant permission for its reporters to cover certain events.

The new Iraqi government's complaints are some of more than 450 formal complaints that the countries of regions have filed to Qatari diplomats. In fact, countries like Saudi Arabia and Algeria do not hesitate to classify the news broadcaster as responsible for the spreading of terrorism in the region. Al Jazeera's height and influence have changed the Jihadist movement's perception of the role that traditional media can play in terrorist strategies.

After 9/11, Al Qaeda was able to see how its ideological message met with a certain amount of comprehension in the mass media.

This circumstance led the group to qualify its previous hostility toward these entities. Along these lines, Bin Laden himself made reference to this "turning point": "Apart from that, there is the group of the media people and writers who have remarkable impact and a big role in directing the battle, and breaking the enemy's morale, and heightening the Ummah's morale (...)

The time has come to have the media take its rightful place, to carry out its required role in confronting this aggressive campaign and the open declared Crusader war by all means that can be seen, heard, and read. It is upon the media people, whether writers, journalists, analysts or correspondents, to exercise responsibility in reporting events, and to carry out their required role by showing the Ummah the reality of the events, and to announce the real intentions of the enemy, to reveal his plans and his tricks."

Numerous individuals have tried to denounce the fact that Al Jazeera's connivances with jihadist terrorism go beyond mutual sympathies or certain ideological principles shared by the two groups. Those that hold this point of view believe in the existence of an explicit agreement on the broadcasting of propagandistic material. They base their theory on information published in 2003 by Al Quds Al Arabi, an Arabic language newspaper that reproduced an interview done on a jihadist on-line forum of "Abi Osama," the alleged head of Al Qaeda's media division. In this interview, the director of the Sahab Institute for Media Production openly recognizes the existence of this link, even stating that "the station is obligated to broadcast any videotape we send to it".

Aside from the amount of credibility that we can give to this information, the fact is that this network's attitude is determined by more obvious and easily perceivable factors: The first of these is of economic nature. Al Jazeera's attitude, like other networks, adjusts to a series of calculations on profitability analysis. One of the principal criticisms made about the broadcaster is the near nonexistence of criticisms of the Qatari government, the chief economic supporter of company. Differing from the way it treats other regimens in the area, it is much more careful when questioning the lack of freedom in its home country and the efficiency of the governing class. Despite its reputation for independence and honesty, this satellite TV station is subject to the same restrictions as any other government channel.

The subservient relationship that it has with its owners (some of the most loyal allies to the U.S. in the region) may seem contradictory with respect to its attitude towards Al Qaeda, but it does help us to understand how economic factors can influence in the way a news broadcaster treats a story. In fact, given the available information, it can be affirmed that everything related to Al Qaeda has been wonderful in a business sense for the news channel. The exclusive material on the bombings of Afghanistan was sold by Al Jazeera for a succulent quantity of money: they sold footage on Bin Laden for 20,000 dollars a minute and even a three minute long video with a 1998 interview of him for 250,000 dollars.

The terrorist organization has become an highly valuable accumulator of resources in a setting in which the majority of Middle Eastern governments have stopped advertising the broadcaster in retaliation for the continual criticisms that they have received. Ibrahim Helal, editor in chief of the channel, recognized this fact in an interview for the BBC: "It is necessary to admit that to have these tapes in our power is a novelty that cannot be rejected from an informative and commercial point of view. I do not believe that any television it had thought two times. On having showed these tapes, we generate a major number of television viewers and sell better."

Beyond the economic motivations, the members are clearly convinced of what the editorial line of the news network should be. In this broadcaster's staff, it is possible to find a heterogeneous cast of professionals from very diverse backgrounds and political and religious orientations: members of the Muslim Brotherhood, westernized journalists, leftist intellectuals, etc. Regardless of these traits, all of them share a strong rejection toward the United States' traditional role in the region; and this is where Al Qaeda finds its place among them. Although it is true that jihadist plans for the world are completely different from what many members of Al Jazeera wish for (some of whom sincerely hope for the democratization and opening of these societies,) both visions require as a previous step the restriction of the West's role in the Islamic world's affairs.

This relates to applying Ronfeld and Arquilla's theory on the "netwars" where participants who "network" receive the benefits of other groups' work, even when they have opposing objectives. In their treatment of information, Al Jazeera uses the same editorial process as other Western media; however, the final product is completely different. During the Iraqi war, they had a clearly sympathetic tone toward insurgent Iraqis and a clearly hostile one toward Americans.

The same thing happened with the Taliban and with the cutting irony used when reporting on Muslim governments that are self-declared allies of the U.S. in the War on Terror. Al Jazeera shows an obsessive willingness to always give "both sides" of the same situation, and this has had contradictory and troublesome effects. On one hand, it has provided Al Qaeda with the same legitimacy and attention that is given to legitimate participants in a war, but on the other hand, it has also broadcast messages completely unknown in the Arabic-Muslim world's audio-visual repertoire. For example, before this network's existed, it was normal for a resident of Muslim country to never have heard an Israeli spokesperson explain his version of the conflict.

It should be recognized, however, that the process involved between receiving and broadcasting terrorist videos and other materials is still not automatic. At times, the network has chosen to only broadcast a certain part of materials received simply state that it has received materials without putting them on the air. In short, Al Qaeda's choosing of Al Jazeera is logical if we take into account the clear interest that the terrorist organization has for diffusing its message to the Muslim world. Bin Laden found in Al Qaeda not only a powerful mechanism of transmission, but also an inclination toward his message and an editorial backing difficult to perceive in other wide reaching mass media.

A member of Al Qaeda recognized that its choice of network was strongly related to its clear and irrefutable history of support of the mujahideen: "Sheikh Osama knows that the media war is not less important than the military war against America. That's why al Qaeda has many media wars. The Sheikh has made al Qaeda's strategy something that all TV stations look for. There are certain criteria for the stations to be able to air our videos, foremost of which is that it has not taken a previous stand against the mujahideen. This maybe explain why prefer Al Jazeera to the rest."

Internet and Approaching the Media Indirectly

Despite the United States and its allies' massive mobilization of resources in the "War on Terrorism" Al Qaeda has been able to continue supplying the mass media with new propaganda. Each new consignment has meant a new symbolic triumph for the terrorist organization, given that each new communiqué demonstrated their capacity to evade their powerful enemies' siege.

The way in which Bin Laden's organization has been capable of maintaining this line of communication was a mystery before some of the most noteworthy members' capture. This was especially true in Abu Faraj al Libbi's case who was arrested in Pakistan in May 2005. His interrogation revealed how Al Qaeda utilized a complicated network of messengers who distributed principal communiqués (with Ayman Al Zawahiri or Bin Laden himself as their protagonists.)

These couriers took anywhere from six to twelve weeks to travel less than 70 miles of intricate routes between the Afghan-Pakistani border and Al Jazeera's office in Islamabad. The messengers (many of whom are recruited among the preachers who travel through this zone by foot,) for security reasons, only travel a small part of the route, being unaware of the origin, the final recipient, and the content of the material that they transported. On occasions, the last stage, instead of involving bringing the message to the TV network, involves an intermediary sending the file by Internet.

This complicated network of links has been challenging for intelligence services that have been left to powerlessly contemplate how Al Qaeda has continued maintaining its propagandistic capacities intact. In this situation, Al Qaeda has been able to appear to the world as durable and resistant, despite the losing its sanctuary in Afghanistan and the death and imprisonment of many members. Pakistani authorities did, however, manage to intercept the sending of these messages in at least two occasions in 2003 and 2004.

This allowed U.S. intelligence services to develop a better understanding of the network that allowed Al Qaeda to keep its propagandistic system active. In fact, the American air attack of Damadola, a Pakistani town, in January 2006 (that resulted in the deaths of important members of the organization and almost reached Ayman

Al Zawahiri) has been attributed to U.S. intelligence's capability to infiltrate this web propaganda distribution.

Al Qaeda understands that the type of relationship that it has with the mass media in recent years highly threatens the organization's and its members' security. Its desire to eliminate these vulnerabilities has lead the terrorist organization put new technology to even more use. This way, Bin Laden's organization has opted to disseminate its most recent news developments on the Internet. This new strategy does not mean that they disregard the chance to have top roles on the mass media's agenda; it means that they work toward an indirect approximation strategy.

In other words, the group continues taking into consideration that a large part of its strategy's success depends on its capacity to reach the mass media; however, it looks to make this contact in a safer and more effective manner. Al Qaeda has learned from the propagandistic experience of other terrorist groups that surround it. Many of them (for example, the group founded by Abu Musab Al Zarqawi o the Saudi faction of Al Qaeda) almost never establish direct contact with the mass media, concentrating their communicative activities in cyberspace.

These methods have prevented them from receiving wide media attention. Paradoxically, the mass media themselves use the web to look for footage and messages that act to further illustrate their news stories. The very existence of these elements on the Internet is a story in and of itself, without the need for other intermediaries. In this way, the news is compelled to reflect and present those events that anonymously have the capability to gain noteworthy repercussions on international public opinion.

Al Qaeda has no problems in adapting to the demands of this new media and has incorporated a series of innovations that tend to obtain the maximum repercussion for its messages. Thus, Al Qaeda's most recent communiqués have been preceded by a series of publicity banners that it puts on jihadist forums advertising the imminent diffusion of these materials. Using this practice, the group manages not only to create expectation among its followers, but also to alert the media so that they eco the new message from the very instant in which this new material is placed on the web.

At times, TV stations compete with each other to be first to broadcast the most recent developments. For Al Qaeda, the Internet is not only method to reach the media in a safer and more immediate way, it also is a turning point in their communication strategy given that the web devalues the importance of traditional media. For the first time in history, cyberspace allows for there to be direct communication between a terrorist and his "public." Terrorists control their messages well, always saying exactly what they want to say and when they want to say it. In the past, directly sending materials to the mass media was quite problematic for a terrorist group.

Firstly, there were chances that the message would be ignored, slanted or even manipulated. Terrorists especially needed to take into account in their calculations what the media was willing to tolerate and what it was not. Thus, for example, the chance to distribute a long and dense ideological discourse was discarded when we take into account that TV stations are characterized by time limitations that affect the contents of the news and encourage them to search for visual effects.

Diffuse a message like this was difficult, including for a legitimate political leader. Secondly, and as we have already indicated, the perpetual sending of materials presented security problems, due to the possibilities that counter-terrorist agencies could follow the path of these messages from the sender to the media. These forced terrorist groups not make frequent contact with this method of diffusion. Furthermore, the repeated sending of material to a determined media could cause public opinion to unavoidably associate a terrorist group with the media that attended to these initiatives. The fact that the media themselves selected the materials that were clearly newsworthy avoided their being prejudiced by the violence that terrorism communicated.

The Internet does only permit the avoidance of the aforementioned limitations, it also has made it possible for the mass media to ignore a series of moral restrictions that highly benefit terrorism strategies. In the past, television was the only means by which terrorist violence could be published on a grand scale. This meant that those who were responsible for this media were the only ones who could decide if the public at large should see material of this nature. But since this type propaganda has been available on the Internet, TV channels feel that they have been released from making this difficult moral decision; they are now not the only ones who are responsible for the viewing public witnessing this cruel and bloody show.

The blurring of this responsibility has caused television stations to not be excessively scrupulous when showing macabre or dramatic footage; they have become terrorism's involuntary accomplices. Al Qaeda's relationship with the media has passed through different phases. Its different perceptions and the way in which this terrorist group has attempted to utilize the mass media are really the results of estimated media impact and have much less to do with ideological or religious interpretations.

Thus, in the initial phases, starting with its appearance in the late 1980s in Afghanistan up until the late 1990s, perception of the group has been shaped Western media (especially the United States',) given its dominance in the world media scene. Al Qaeda shares its hostility toward the media with other terrorist groups holding them responsible for hiding or distorting its message. But, it is also conscious of the importance that these channels have for reaching a wide audience and, with these methods, achieving its ultimate goal: worldwide Islamic revolt. Terrorism sees mass media, especially TV, as having a series of characteristics that make them especially vulnerable to terrorism's attempts to monopolize the "public sphere."

The second phase is related to the Arabic television network Al Jazeera starting with its appearance in 1996. This is time when Muslim controlled press starts to gain prominence. Al Qaeda detects new opportunities, characterized by greater repercussions for its message and the existence of certain channels willing to interpret reality from a more jihadist ideology friendly perspective. The terrorist group starts a relationship not free of certain complicity with Al Jazeera, a network that also has an enormous influence in the Muslim world; this network becomes a principal objective in Al Qaeda's communication strategy.

The current phase corresponds to the Internet becoming generalized as part the mass media. Certain jihadist groups exploit it following the invasion of Iraq in 2003. The Internet not only allows them to avoid certain operational risks, it also allows them to gain access to traditional media. Thus, the old "terrorist dream" of being able to establish direct contact between the group and a potentially unlimited public comes true.

Bibliography

Adams, James. *New Spies: Exploring the Frontiers of Espionage.* London: Pimlico, 1995.

Adams, James. *Next World War: Computers Are the Weapons and the Front Line is Everywhere.* New York: Simon & Schuster, 1998.

Adams, Neal. *Terrorism & Oil.* Tulsa, OK: PennWell, 2003.

Akhavan, Jacqueline. *Chemistry of Explosives.* Cambridge, MA: Royal Society of Chemistry Information Service, 1998.

Alali, A. Odasuo. *Media Coverage of Terrorism: Methods of Diffusion.* Terrorism and the Mass Media. Newbury Park, CA: Sage Publications, 1991.

Alali, A. Odasuo. *Terrorism and the News Media: A Selected Annotated Bibliography.* Jefferson, NC: McFarland & Company, 1994.

Alexander, Dean C. *Business Confronts Terrorism: Risks and Responses.* Madison, WI: University of Wisconsin Press, 2004.

Alexander, Dean C. *Terrorism and Business: The Impact of September 11, 2001.* Ardsley, NY: Transnational, 2002.

Alexander, Yonah. *Control of Terrorism: International Documents.* New York: Crane Russak, 1979.

Alexander, Yonah. *Cyber Terrorism and Information Warfare: Threats and Responses.* Terrorism Documents of International Local Control. Ardsley, NY: Transnational Publishers, 2001.

Alexander, Yonah. *Future Terrorism Trends.* Washington, DC: Global Affairs, 1991.

Alexander, Yonah. *In the Camera's Eye: News Coverage of Terrorist Events.* TLBS Brassey's Terrorism Library. Washington, DC: Brassey's, 1990.

Alexander, Yonah. *Political Terrorism and Energy: The Threat and Response.* New York: Praeger, 1982.

Alexander, Yonah. *Terrorism: Future Trends.* Washington, DC: Global Affairs, 1991.

Alexander, Yonah; Latter, Richard, editors, *Terrorism and the Media: Dilemmas for Government, Journalists and the Public.* Washington, DC: Brassey's, 1990.

Anti-Defamation League, *"Propaganda of the Deed": The Far Right's Desperate 'Revolution'.* New York: Antidefamation League, 1985.

Anti-Defamation League, *High-Tech Hate: Extremist Use of the Internet.* New York: Anti-Defamation League, 1997.

Apikyan, Samuel. *Countering Nuclear and Radiological Terrorism.* Brussels, Belgium: Springer, 2006.

Arquilla, John. *Networks and Netwars: The Future of Terror, Crime, and Militancy.* Santa Monica, CA: RAND, 2001.

Ashwood, Thomas M. *Terror in the Skies.* New York: Stein and Day, 1987.

Avruch, Kevin. *Culture and Conflict Resolution.* Washington, DC: United States Institute of Peace, 1998.

Baker, David. *Biological, Nuclear, and Chemical Weapons.* Vero Beach, FL: Rourke, 2006. Juvenile Audience.

Ballard, James David. *Terrorism, Media and Public Policy: The Oklahoma City Bombing.* Hampton Press communication series. Cresskill, NJ: Hampton Press, 2005.

Balmer, Brian. *Britain and Biological Warfare: Expert Advice and Science Policy, 1930-65.* New York: Palgrave, 2001.

Baudrillard, Jean. *Spirit of Terrorism and Requiem for the Twin Towers*. London: Verso, 2002.

Baudrillard, Jean. *Spirit of Terrorism, and Other Essays*. New York: Verso Books, 2003.

Cashman, John R. *Emergency Response to Chemical and Biological Agents*. New York: Lewis, 2000.

Casil, Amy Sterling. *Coping With Terrorism*. New York: Rosen Publishing, 2004. Juvenile Audience.

CBS News Productions, *Terrorism in Our World*. 4 videocassettes (VHS). Wynnewood, PA: Schlessinger Media, 2003. Juvenile Audience.

Dando, Malcolm. *Preventing Biological Warfare: The Failure of American Leadership*. New York: Palgrave, 2002.

Das, Dilip. *Meeting the Challenges of Global Terrorism: Prevention, Control, and Recovery*. Lanham, MD: Rowman & Littlefield, 2003.

Devji, Faisal. *Landscapes of the Jihad: Militancy, Morality, Modernity*. Ithaca, NY: Cornell University Press, 2005.

Dewar, Michael. *Weapons and Equipment of Counter-Terrorism*. 2nd ed. New York: Sterling, 1995.

El Mahdy, Galal. *Disaster Management in Telecommunications, Broadcasting, and Computer Systems*. New York: John Wiley, 2001.

Ellis, John. *Police Analysis and Planning for Chemical Biological and Radiological Attacks*. Springfield, IL: C.C. Thomas, 1999.

Evans, Ernest. *Calling a Truce to Terror: The American Response to International Terrorism*. Westport, CT: Greenwood Press, 1979.

Gal-Or, Noemi. *International Cooperation to Suppress Terrorism*. New York: St. Martins Press, 1985.

Gal-Or, Noemi. *Tolerating Terrorism in the West*. New

York: Routledge, 1991.

Gambill, Andrea. *No One Should See What I Have Seen: A Book for Those Who Have Experienced Terrifying and Horrific Tragedy*. Omaha, NE: Centering Corporation, 2002.

Ganor, Boaz, editor. *Post-Modern Terrorism: Trends, Scenarios, and Future Threats*. Herzliya, Israel: International Policy Institute for Counter-Terrorism, 2005.

Garbutt, Paul E. *Assassin: From Lincoln to Rajiv Gandhi*. Shepperton: Ian Allan, 1992.

Garcia, Mary Lynn. *The Design and Evaluation of Physical Protection Systems*. Boston, MA: Butterworth Heinemann, 2001.

Gardela, Karen. *RAND Chronology of International Terrorism 1986*. Santa Monica, CA: RAND, 1990.

Gardela, Karen. *RAND Chronology of International Terrorism 1987*. Santa Monica, CA: RAND, 1991.

Ganor, Boaz. *The Counter-Terrorism Puzzle: A Guide for Decision Makers*. New Brunswick, NJ: Transaction Publishers, 2005.

Gardela, Karen. *RAND Chronology of International Terrorism 1988*. Santa Monica, CA: RAND, 1992.

Gardell, Mattias. *Gods of the Blood: The Pagan Revival and White Separatism*. Durham, NC: Duke University Press, 2003.

Gareau, Frederick H. *State Terrorism and the United States: From Counterinsurgency to the War on Terrorism*. Atlanta, GA: Clarity Press, Inc., 2004.

Garfinkle, Adam. *Practical Guide to Winning the War on Terrorism*. Stanford, CA: Hoover Institution Press, 2004.

Gay, Kathlyn. *Encyclopedia of Political Anarchy*. Santa Barbara, CA: ABC-CLIO, 1999.

Gay, Kathlyn. *Militias: Armed and Dangerous*. Springfield, NJ: Enslow Publishers, Inc., 1997. Juvenile Audience.

Gay, Kathlyn. *Silent Death: The Threat of Chemical and Biological Terrorism*. Brookfield, CT: Twentyfirst Century Books, 2001. Juvenile Audience.

Gearty, Conner. *Terrorism*. London: Phoenix, 1997.

Gearty, Conor. *The Future of Terrorism*. London: Phoenix, 1997.

Geissler, Erhard. *Biological and Toxin Weapons: Research, Development, and Use From the Middle Ages to 1945*. New York: Oxford University Press, 1999.

Gerstein, Daniel M. *Security America's Future: National Strategy in the Information Age*. Westport, CT: Praeger, 2005.

Ghosh, Tushar K. *Science and Technology of Terrorism and Counterterrorism*. New York: Dekker, 2002.

Gilbert, Paul. *New Terror New Wars*. Washington, DC: Georgetown University Press, 2003.

Hanle, Donald J. *Terrorism: The Newest Face of Warfare*. Elmsford, NY: Pergamon-Brassey, 1989.

Hansen, Jon. *Oklahoma Rescue*. New York: Ballantine Books, 1995.

Harclerode, Peter. *Secret Soldiers: Special Forces in the War Against Terrorism*. London: Cassell, 2000.

Hoffman, Bruce. *Future Trends in Terrorist Targeting and Tactics*. Santa Monica, CA: RAND, 1993.

Hoffman, Bruce. *Future Trends of Terrorist Marketing*. Santa Monica, CA: RAND, 1993.

Howitt, Arnold. *Countering Terrorism: Dimensions of Preparedness*. Cambridge, MA: MIT Press, 2003.

Jenkins, Brian M. *High Technology Terrorism and Surrogate War: The Impact of New Technology on Low-Level Violence*. Santa Monica, CA: RAND, 1975.

MacDonald, Donald. *Parliaments Against Terrorism*. Ottawa, Canada: Library of Parliament, 1986.

MacDonald, Eileen. *Shoot the Women First.* New York: Random House, 1991.

Maley, William. *Fundamentalism Reborn : Afghanistan and the Taliban.* New York: New York University Press, 2001.

Maniscalco, Paul M. *Understanding Terrorism and Managing the Consequences.* Upper Saddle River, NJ: Prentice Hall, 2002.

Mann, Michael. *Dark Side of Democracy: Explaining Ethnic Cleansing.* New York: Cambridge University Press, 2004.

Manwaring, Max G. *The Search for Security: A U.S. Grand Strategy for the Twenty-First Century.* Westport, CT: Praeger, 2003.

Mockaitis, Thomas R. *Grand Strategy in the War Against Terrorism.* Portland, OR: Frank Cass, 2003.

Moghaddam, Fathali. *Understanding Terrorism: Psychosocial Roots, Consequences, and Interventions.* Washington, DC: American Psychological Association, 2003.

Norris, John. *NBC: Nuclear, Biological and Chemical Warfare on the Modern Battlefield.* London: Brassey's, 1997.

Norton, Augustus R. *Understanding the Nuclear Terrorism Problem.* Gaithersburg, MD: International Association of Chiefs of Police, 1979.

Nusse, Andrea. *Muslim Palestine: The Ideology of Hamas.* Amsterdam: Harwood Academic Publishers, 1998.

Nyatepe-Coo, Akorlie A. *Understanding Terrorism: Threats in an Uncertain World.* Upper Saddle River, NJ: Pearson, 2004.

Paletz, David L. *Terrorism and the Media.* Newbury Park, CA: Sage, 1992.

Schechter, Danny. *Media Wars: News at a Time of Terror.* Lanham, MD: Rowman & Littlefield, 2003.

Taillon, J. Paul. *Evolution of Special Forces in Counter-*

terrorism: The British and American Experiences. Westport, CT: Praeger, 2001.

Unger, Rhoda Kesler. *Terrorism and Its Consequences.* Malden, MA: Blackwell, 2002.

United Nations, *International Instruments Related to the Prevention and Suppression of International Terrorism.* New York: United Nations, 2001.

Van Der Veer, Peter. *Media, War, and Terrorism: Responses from the Middle East and Asia.* New York: Curzon, 2004.

Vohryzek-Bolden, Miki. *Domestic Terrorism and Incident Management: Issues and Tactics.* Springfield, IL: C.C. Thomas, 2001.

Waldron, Jonathan K. *Maritime Security Initiatives: Implementing the New Regulations.* Rockville, MD: ABS Consulting, 2004.

Ward, Richard H. *Terrorism and the New World Disorder.* Chicago, IL: Office of International Criminal Justice, University of Illinois at Chicago, 1998.

Webster, William H. *Wild Atom: Nuclear Terrorism.* Washington, DC: Center for Strategic and International Studies, 1998.

Wecht, Cyril H. *Forensic Aspects of Chemical and Biological Terrorism.* Tucson, AZ: Lawyers and Judges Pub, 2004.

Weimann, Gabriel. *Theater of Terror: Mass Media and International Terrorism.* New York: Longman, 1994.

Weintraub, Pamela. *Bioterrorism: How to Survive the 25 Most Dangerous Biological Weapons.* New York: Citadel Press, 2002.

Wilcox, David. *Domestic Preparedness and the WMD Paradigm.* Washington, DC: U.S. Army Command, 1998.

Wilkinson, Paul. *Contemporary Research on Terrorism.* Aberdeen: Aberdeen University Press, 1987.

Index